TELECOM MANAGEMENT CRASH COURSE

Other McGraw-Hill Telecommunications Crash Courses

Austin	*Telecom Security Crash Course*
Bedell	*Wireless Crash Course*
Kikta & Fisher	*Wireless Internet Crash Course*
Louis	*Broadband Crash Course*
Louis	*M-Commerce Crash Course*
Shepard	*Optical Networking Crash Course*
Shepard	*Telecom Crash Course*
Shepard	*Telecommunications Convergence,* 2/e
Vacca	*i-mode Crash Course*

TELECOM MANAGEMENT CRASH COURSE

Managing and Selling Telecom Services and Products

P. J. LOUIS

McGraw-Hill
New York Chicago San Francisco Lisbon
London Madrid Mexico City Milan New Delhi
San Juan Seoul Singapore Sydney Toronto

Library of Congress Cataloging-in-Publication Data

Louis, P. J.
Telecom management crash course : managing and selling telecom services and products / P. J. Louis.
p. cm.
ISBN 0-07-138620-3
1. Telecommunication—Management. I. Title.

HE7661.L68 2002
384'.068—dc21

2002066042

McGraw-Hill

A Division of The McGraw-Hill Companies

Copyright © 2002 by P. J. Louis. All rights reserved. Printed in the United States of America. Except as permitted under the United States Copyright Act of 1976, no part of this publication may be reproduced or distributed in any form or by any means, or stored in a data base or retrieval system, without the prior written permission of the publisher.

1 2 3 4 5 6 7 8 9 0 DOC/DOC 0 8 7 6 5 4 3 2

ISBN 0-07-138620-3

The sponsoring editor for this book was Stephen S. Chapman and the production supervisor was Pamela A. Pelton. It was set in Fairfield Medium by Patricia Wallenburg.

Printed and bound by R. R. Donnelley & Sons Company.

McGraw-Hill books are available at special quantity discounts to use as premiums and sales promotions, or for use in corporate training programs. For more information, please write to the Director of Special Sales, McGraw-Hill Professional, Two Penn Plaza, New York, NY 10121-2298. Or contact your local bookstore.

This book is printed on recycled, acid-free paper containing a minimum of 50 percent recycled, de-inked fiber.

Information contained in this book has been obtained by The McGraw-Hill Companies, Inc. ("McGraw-Hill") from sources believed to be reliable. However, neither McGraw-Hill nor its authors guarantee the accuracy or completeness of any information published herein, and neither McGraw-Hill nor its authors shall be responsible for any errors, omissions, or damages arising out of use of this information. This work is published with the understanding that McGraw-Hill and its authors are supplying information, but are not attempting to render engineering or other professional services. If such services are required, the assistance of an appropriate professional should be sought.

DEDICATIONS

Dedicated to my darling wife, Donna, and our children, Eric and Scott. Their love, tolerance, patience, and support have enabled me to achieve my professional goals.

In memory of my late parents, Richard Louis and Jennie Chin Louis.

In memory of the late Harry E. Young; my friend, superior, and mentor.

CONTENTS

PREFACE

In 1996, the Telecom Act was passed in the United States. On a global basis, nations were also privatizing their telephone companies, selling spectrum, and opening their telecom markets for more competition. Since 1996, there have been numerous changes in the telecom business, including the aggressive marketing of university-provided postgraduate courses on telecommunications, shifts in the employment market, and aggressive private investing. However, the most important, yet subtle, change has been the world's perception of telecommunications employment. Where once, working for the telephone company was considered a haven for long-term employment, now suddenly, working for the telephone company and in other areas of telecommunications is considered "the place to be."

The breakup of the Bell System and the privatization of other national Public Telegraph & Telephone (PTT) companies opened the employment market to the world. The emergence of wireless communications as the current telecom growth sector has greatly contributed to the pervasive interest in telecom in the financial sector and to the creation of jobs. The birth of the Internet as a commercial product has also made a vast contribution to the shifts in people's attitudes towards telecom. Despite the recent downturn in the marketplace, the telecom sector has captured the imagination and attention of everyone who believes it is possible to gain some kind of financial reward there from.

Despite the market downturn, the demand for telecommunications services will continue to grow. Telecommunications is a basic social infrastructure requirement like electricity and water. Telecom employment went from being a public utility job to a profession. There are people who used to call themselves telephone company or cellular carrier workers; now they call themselves telecom professionals: The shifts in the job market have created a new kind of professional; the telecom professional. This is a broad term to describe someone who has gathered a variety of skills in one area of expertise, with many different ancillary skills. The telecom manager is a professional focused on the overall management of a telecom company.

INTRODUCTION

This is a business and management how-to book about the basics of running a telecommunications company. It takes a functional business look at the fundamentals of how hardware manufacturing companies, software development companies, and service providers operate in the telecommunications business space. This is not a book of theory, but rather one based on my experiences and the experiences of my colleagues. It can and occasionally will quote theory, but will lean toward using simple and understandable terms.

This book categorizes any telecom company as belonging to one of two groups: service providers (also known as carriers) or vendors.

The book is about how telecom companies; are generally organized and managed. It addresses sales and marketing fundamentals, describes the functions of various organizations, the role of regulation, evaluating the marketplace, the financial value of products and services, customer care, billing, technology, and the role of the customer. Running a telecom company is not just a simple matter of "building it" and hoping customers will come, it is about "building the right thing to sell it to the customer."

This book avoids doing case studies that look at how specific companies succeeded or failed because too often, I have seen professionals ignore any advice or insight when a specific example is given. Professionals will often say that company XYZ made widgets and they are in the business of making ball bearings; since the companies are different, any lessons to be learned are not applicable to them. My experience indicates that speaking in generalities is usually enough to get fellow telecom professionals to do their own self-analysis (in silence and the privacy of their own thoughts). Remember the old leadership adages, "never ever show the other person you're sweating" and "don't show the other person how stupid you are."

The tragedy of working over the long term in any profession is the belief that one "has been there and done it" and therefore need not learn anything or find anything new to learn. My advice to people in this position is to, "wake up." Unemployment is the single greatest equalizer for telecom professionals with the indestructible/indispensable attitude. The telecom professional lives in a world where the average employment span in a job can be measured in months, not years. Telecom is not the only profession suffering from the new global economy, so the situation just outlined should not be difficult to comprehend. Understanding the mechanics of a telecom company ought to be a concern for anyone working or investing in the telecom industry.

Those who began investing (speculating) in the telecom business space already know that telecom has its own language and its own way of operating. However, telecom is like any other industry: it has its language and practices but the industry operates on the same metrics as any other business. The goal for a telecom company is to make "money" by selling its services or products. The big question that has not been asked by most investors is: Is this company being managed well?

Business majors in universities will find this book covers basic business and management theories, though everything is in "plain speak." This is not a textbook, but a working guide to understanding the telecom business from a management perspective. One should read this book with an eye toward continually learning about the telecommunications business. What follows is a summary of the material to be considered.

CHAPTER 1: WHAT IS A TELECOMMUNICATIONS COMPANY?

What kinds of telecom companies are there? How are the companies organized? Is there a generic organizational structure? What kinds of organizations are there? What is the best organizational structure to use? Is there a difference in vendor companies? Is there a difference in how vendors work?

CHAPTER 2: ORGANIZATIONAL STRUCTURES

What comprises a company? What kinds of organizational structures are used? We will find sets of basic organizational structures that can be used by any company. The structure used by a corporation depends on the type of product or service the telecom company is providing. How well the organization functions is dependent on the quality of the management team.

CHAPTER 3: THE TELECOMMUNICATIONS SERVICE PROVIDER

The service provider telecommunications services to the user. It is not a supplier of equipment, even though providers like cellular carriers sell handsets. A service provider's core function is to enable the connection and transfer of information between users. Telecom service providers take responsibility for the physical transport of information between users. This responsibility also includes data integrity of the information upon its arrival at the destination.

The many different kinds of service providers and users of telecom services can be classified into basic business categories: retail and wholesale. Technical categories include telephony wireline, wireless, and Internet. The Internet is really a new medium toward which all industry segments are converging.

Some telecom carriers are centrally organized, with a powerful headquarters. Others empower their regions to act independently of the headquarters and rely very little on headquarters for direction. Integrating technology and business objectives is a challenge for carriers in the current environment of intense competition.

CHAPTER 4: THE TELECOM VENDOR

The telecom vendor today faces an economy that is global, rapidly changing, replete with competitors, filled with spoilers hoping to be bought up by competitors, regulated to some extent, and containing customers who suddenly go out of business; it is litigious, has small market windows, and requires fast times to market.

The business environment has become so complicated for vendors that, rather than becoming suppliers of all kinds of telecom hardware and software, they have become focused on being successful in a very small niche. If a vendor is successful, it might expand its product line for long-term viability. Various management skills have been required for those vendors able to ride out the rough economic storms of 2000–2001. In fact, during the writing of this book, the "successful vendors" have been going through their own periods of distress.

The integration of technology and business objectives has been an extremely (and understandably) difficult goal for all telecom companies.

CHAPTER 5: STRATEGY, TACTICS, AND SHARED VALUES

Strategic planning sets the direction of the company. How does a team of highly skilled and diverse managers set the company's course? The creation of a plan for the future of a company is not just merely an intellectual exercise. Employees, investors, and customers are affected by the changes created by management. Some of these changes in direction are deliberate and some are the result of circumstance.

CHAPTER 6: MARKET FOCUS—UNDERSTANDING THE TELECOMMUNICATIONS MARKETS

The concept of a company being sales driven rather than customer driven is an ongoing struggle, which ultimately affects the entire company. One approach focuses on selling the customer what the company makes. The other focuses on meeting the needs of the customer. The need to ensure a tight integration of sales, marketing, product delivery, and product development efforts is of paramount importance. We call this a market focus. Many startup companies have failed because the left hand did not know or even care what the right hand was doing.

Another challenge is creating a market for a product where there had been none. There is a structured planning approach that can be used to achieve this end.

CHAPTER 7: DEVELOPING MARKET DRIVEN STRATEGIES—HOW?

In the late 1990s through 2000, the telecommunications industry underwent an unprecedented period of enormous growth. Unfortunately, the industry focused more on spending money and building big networks than on building what the market could use. This chapter is a continuation of Chapter 5, taking a deeper look into the processes involved in developing market strategies.

Keeping a management team focused on the goals of the company is not as simple as it sounds. Reconciling the personality issues involved in developing, agreeing on, and executing any strategy can be an impossible task.

CHAPTER 8: BUILDING THE MANAGEMENT TEAM AND BUILDING THE MARKET-DRIVEN COMPANY

This chapter focuses on the issues surrounding the creation of a winning management team and the building the market-driven company.

The book concludes with a glossary of useful terms.

ACKNOWLEDGMENTS

I have had the pleasure of working for a number of great managers and leaders who took the time to show me "the ropes." I want to thank them: Charles P. Eifinger, the late Howard Schuster, the late Carl Ripa, Lawrence J. Chu, the late Harry Young, the late Jerry Pfeiffer, Walter T. Gorman, Richard Murphy, Alexander Korn, Richard Robben, Nancy Anderson, Robert Mandell, and my father. Their combined experience adds up to almost 500 years—they were great teachers.

To all my colleagues who have influenced me through our interactions and debates.

My illustrator, Melissa N. Brown.

CHAPTER ONE

WHAT IS A TELECOMMUNICATIONS COMPANY?

There are a number of company types in the telecommunications space, including:

- Makers of hardware
- Makers of software
- Providers of service to the hardware and software companies
- Providers of service (carrier/service provider) to the user
- Providers of service to other providers of service

Because telecommunications equipment is complex, these categories can be further subdivided. The corollary is that the business of creating the product is just as complex. Service provisioning is as complex a business as one can be in, especially since the service is so technology dependent. Technology is constantly changing and evolving and change is therefore a component of the telecommunications business.

MAKERS OF HARDWARE

Hardware is a generic term used by the telecom professional to refer to all the physical equipment made or assembled by a telecommunications company. Hardware can be anything; it can include: switches, chips, routers, wire, insulation, digital signal processors, generators, batteries, tools, ladders, CD-ROMs, CD-ROM drives, HVAC (heating, ventilation, and air conditioning); magnetic tape (mag tape), cable racks, conduit, CRTs (cathode ray tubes, now typically known as monitors), databases, all the components used to make a switch, router or server, and/or all the material used to construct individual components.

Telecommunications companies in the hardware business are categorized as system integrators, tool providers (usually also manufacturers of the tools), or component manufacturers. Telecommunications service providers usually do business with systems integrators and tool providers. They prefer to do business with the integrators rather than the component manufacturers, because the job of a telecommunications service provider is to provide service, not to make or assemble equipment. As in any business, there is a financial motivation for the way the telecom hardware business is organized around the buying habits of the customer (in this case the telecommunications service provider). This service provider prefers to do business with as few equipment sellers as possible. The perfectly logical set of reasons for buying equipment this way will be discussed later in this chapter.

Tool providers/manufacturers, including companies that make equipment needed to enable the assembly of a system (e.g., a switch or router), the repair of the system, or maintenance of the system, sell to both system integrators and telecom service providers. Tools include screw drivers, crimp tools, wire cutters, splicing tools, ladders, tie wraps, work benches, work gloves, etc. The tool companies have the benefit of being able to sell to many different types of industries, since the telecom community is not the only user of such tools. Other users of common tool sets are the construction, defense, space, and automobile industries.

Figure 1.1 is an illustration of a telecom hardware company.

FIGURE 1.1 Telecom hardware company.

MAKERS OF SOFTWARE

Like hardware, there are different kinds of software. Software can be defined as a set of instructions given to a computer or some piece of hardware. Software can be grouped into two basic categories: source code and object code.

Source code is a set of written instructions to the computer, the language used by programmers. Game programs are a type of source code. These instructions are compiled (translated) into a binary code language called a *machine language*. The machine language is the native (internal) language of the computer hardware and is also known as the *object code*.

Makers of software in the telecom arena sell to terminal device manufacturers, users, and service providers. A terminal device includes any device with which a user directly interacts, including mobile handsets, desktop computers, laptop computers, handheld personal data assistants, and pagers. A user typically purchases a terminal device. The wireless service provider will sell terminal devices to its users as part of the service it is selling. The service provider also purchases software to operate its switching or routing equipment and to operate "back-office" systems such as customer billing, network element management, and network monitoring. Figure 1.2 is an illustration of the software company.

Telecom software is used to support the creation of user services and is under the direct control of the service provider. User services, often called *vertical services* or *custom-calling features*, encompass user features such as caller ID or call forwarding.

PROVIDERS OF SERVICE TO THE HARDWARE AND SOFTWARE COMPANIES

The hardware and software sectors of the telecommunications business often reach outside their companies to obtain a variety of development resources that normally are not kept on staff. This resource gap is filled by consultants and develop-

- Application software
- Handset software
- Office systems software
- Desktop operating system software
- Website developers

FIGURE 1.2 Telecom software company.

ment companies, generically known as development houses. I use the term "vendor" interchangeably with the word "manufacturer." These companies may even fill what is known as the *enterprise space*. The generic term for companies that provide these kinds of services externally is *outsourcing*. We will delve deeper into the concept of outsourcing later in this book.

Boiled down to fundamental concepts instead of market hype, the enterprise business space refers to a company's internal network. Providing services to companies falls into the same categories as providing hardware or software to public telecommunications service providers. Services rendered can include:

- Hardware design
- Software development
- Human resources—many companies now hire external firms to manage their human resource departmental needs
- Marketing consultation
- Hardware and software testing
- System integration (system assembly)
- Warehousing
- Delivery
- Technology consultation—if the company does not possess the expertise, it will either hire someone who does or hire a trainer to teach the technology to the existing staff.

A company that creates hardware or software will often outsource as much as possible and focus on its core specialty. There is always a financial imperative when considering outsourcing. Sometimes companies that render services to manufacturers also provide services to telecommunications carriers.

Intellectual Property

As important as intellectual property is, vendors of hardware and software are still businesses that must generate revenue and net profit. When an external company receives a contract to support a hardware or software vendor, that vendor runs the risk of exposing its core technology (hardware and/or software) to competitors. *Non-disclosure agreements* are always signed between external companies and vendors. However, they are no guarantee of protection.

The only steps to take to protect intellectual property are:

- File a patent to claim ownership of the technology, software, or process.
- File copyright to protect the software.
- If the process is sufficiently complex, filing a patent may not be a good idea. Instead, trade secret protection is the path to follow. There is a famous soft drink whose formula is known only to a handful of key personnel.
- Expose the outside vendor only to that part of the technology or process it needs to see.

When any company hires "outsiders" to provide service, it runs the risk of exposing key aspects, processes, or technology. However, financial resources are always a limiting factor. Gone are the days when a single company could afford to perform every aspect of technology development. Not only are investors wary of investing so much, but companies must and should focus on their core expertise. There will be more on the decision-making process of development later in this book.

PROVIDERS OF SERVICE TO THE USER

A provider of service (to the user) is also generically called a *carrier* or *service provider*. The user can be an individual or a company. A user purchases telecommunications service from a carrier/service provider. The varieties of carriers include:

- Wireline telephone companies
- Cellular (wireless) carriers

- PCS (Personal Communications Service) carriers
- Paging companies
- Cable television companies
- Satellite carriers
- Internet service providers

WIRELINE TELEPHONE COMPANIES

A wireline carrier is your telephone company, both local and long distance. I have classified Local Exchange Carriers (LECs), Competitive Local Exchange Carriers (CLECs), Interexchange Carriers (IXCs or ICs), and Competitive Access Providers (CAPs) as wireline carriers. The regulatory classification called the ILEC (Incumbent Local Exchange Carrier) can also be loosely defined as a wireline carrier. The wireline carrier is any carrier that it is not a wireless carrier, which includes the cellular, PCS, and paging companies.

The wireline carrier classification also covers the Public Switched Telephone Network (PSTN). The term PSTN originally referred to AT&T, the 22 Baby Bells, and all the wireline independent telephone companies. In 1984, at the time of the AT&T divestiture, the long distance carriers were not considered a part of the PSTN. Over the years, with regulation reconfiguring the landscape, the use of the term PSTN has been relaxed. Figure 1.3 is an illustration of a wireline telephone company.

CELLULAR (WIRELESS) CARRIER

The term wireless carrier typically refers to cellular and Personal Communications Service (PCS) carriers. As far as the consumer market is concerned, the only difference is that both provide analog and digital service. However, in the case of digital, the consumer typically understands that the carriers use different kinds of digital technology. In reality, the consumer does not care what technology is behind the wireless service. The consumer's measures of quality are:

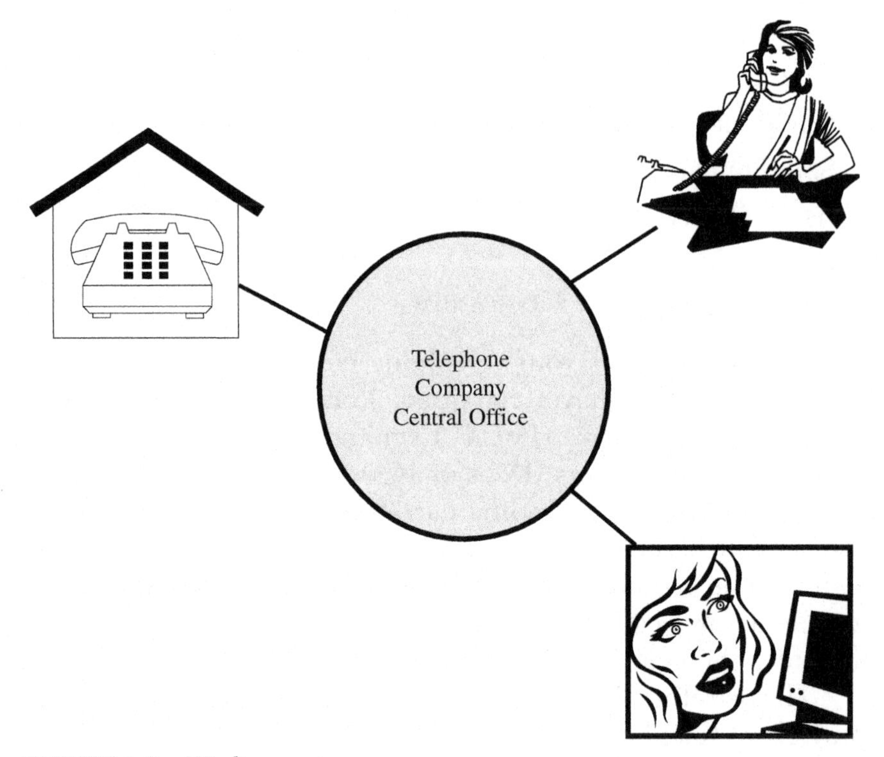

FIGURE 1.3 Wireline carrier.

- Can the parties hear one another without noise or static?
- Can the parties make calls without having to retry the call?
- Can the parties receive calls without going to voice mail first?
- Can the parties get service (as opposed to coverage) wherever they travel? (Coverage refers to radio energy reception, while service means being able to make and receive calls.)
- Is the customer paying a low rate for high-quality service?

Wireless also includes paging companies, LMDS, LMCS, and satellite companies. Local Multipoint Distribution Service (LMDS) in the United States and Local Multipoint Communications Service (LMCS) in Canada are Wireless Local Loop (WLL) technology and business initiatives. LMDS and LMCS carriers operate in the 24 GHz–34 GHz frequency bands. The spectrum had been occupied by the commercial industry and military, in both countries, therefore the licenses are scattered throughout the bands. This chapter will briefly consider these other forms of wireless.

PERSONAL COMMUNICATIONS SERVICES (PCS) COMPANIES/CARRIERS

CELLULAR CARRIERS. In 1946, two-way mobile radio service was introduced. Soon after its introduction, its disadvantages and weaknesses became apparent. From a customer's perspective, there was the issue of competing for RF (radio frequency) channels as well as operational issues due to interference. From an engineer's perspective, the same issues existed. The technical challenge was how to go about giving subscribers a larger pool of RF channels to make their calls and how to reduce interference between subscribers. The quick and simple solution could have been to ignore giving subscribers more RF channels and to physically separate the radio coverage areas to ensure no overlap. Fortunately, no engineer took that approach.

Instead, engineers from Bell Telephone Laboratories began to explore a concept that would reuse frequencies in small radio-coverage areas. These coverage areas (called *cells*) would be linked together using a switch that would enable calls to be made while the caller was moving. Time had to pass and computer/switching technology had to improve before mobile radio service became commercially viable. Availability was another issue, exacerbated by regulatory delays.

Cellular service became commercially available in the United States in 1983. Today there are two basic radio technologies available in the cellular world: analog and digital cellular.

There are multiple types of digital service but these will not be covered in this book. Information on the various kinds of digital technologies available can be found in my other books; *Telecommunications Internetworking*, *M-Commerce Crash Course*, and *Broadband Crash Course*. Detailed explanations of analog cellular service can also be found in these books.

Figure 1.4 depicts a cellular carrier.

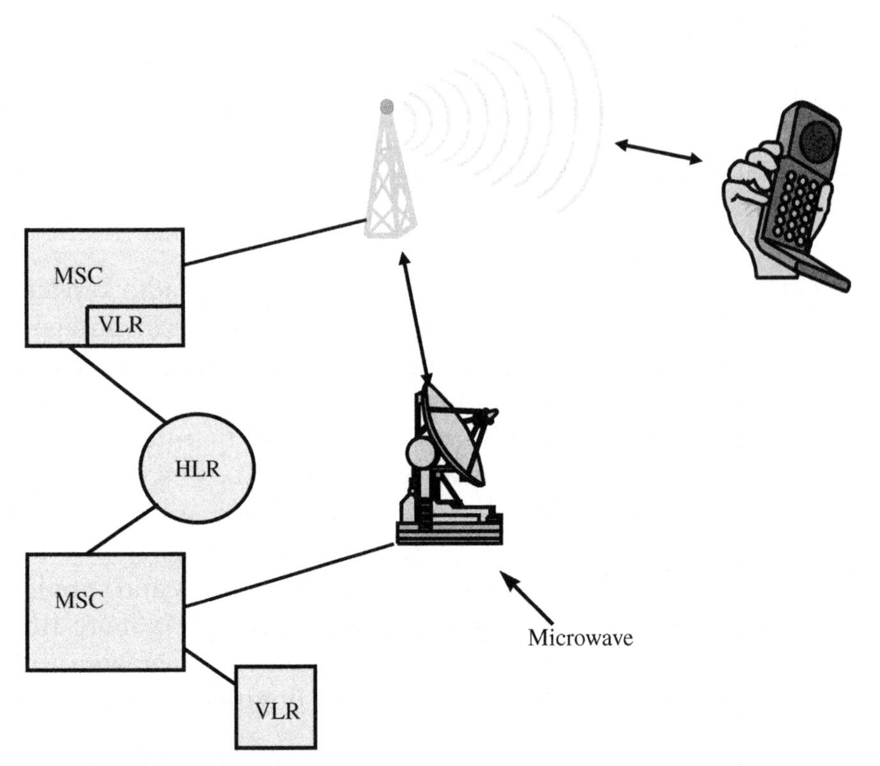

FIGURE 1.4 Cellular carrier.

Cellular carriers in North America operate in the frequency range of 800 to 900 MHz. Carriers in Europe operate in the frequency range from 900 to 1000 MHz.

PERSONAL COMMUNICATIONS SERVICES (PCS) CARRIER. The term Personal Communications Services (PCS) originally envisioned a broad range of services designed to allow people access to the public switched telephone network (PSTN) regardless of their physical location. Today most people believe PCS to be digital cellular, something better than cellular, or cellular without the old baggage (legacy equipment). From a technical perspective, the PCS carrier operates in a frequency band different from that of the cellular carrier. In North America, the PCS carrier operates in the 1800 to 1990 MHz frequency band. In Europe, the PCS carrier operates in the 1900 to 1990 MHz band.

In the mid-1990s many envisioned PCS as having features that supported personal terminals and service mobility. PCS was supposed to combine the many emerging "intelligent network" capabilities of public networks (CCS, ISDN, and AIN) with sophisticated wireless access technologies and related radio network mobility control capabilities. Figure 1.5 is an illustration of a PCS carrier network.

From a technical and conceptual standpoint, there is not much difference between cellular and PCS. The same engineering practices and traffic design techniques are used. Some of the challenges faced by PCS are:

- PCS must contend with high capital costs for equipment and deployment.
- PCS faced competition on day one. Cellular carriers, in their infancy, did not have much competition to worry about.

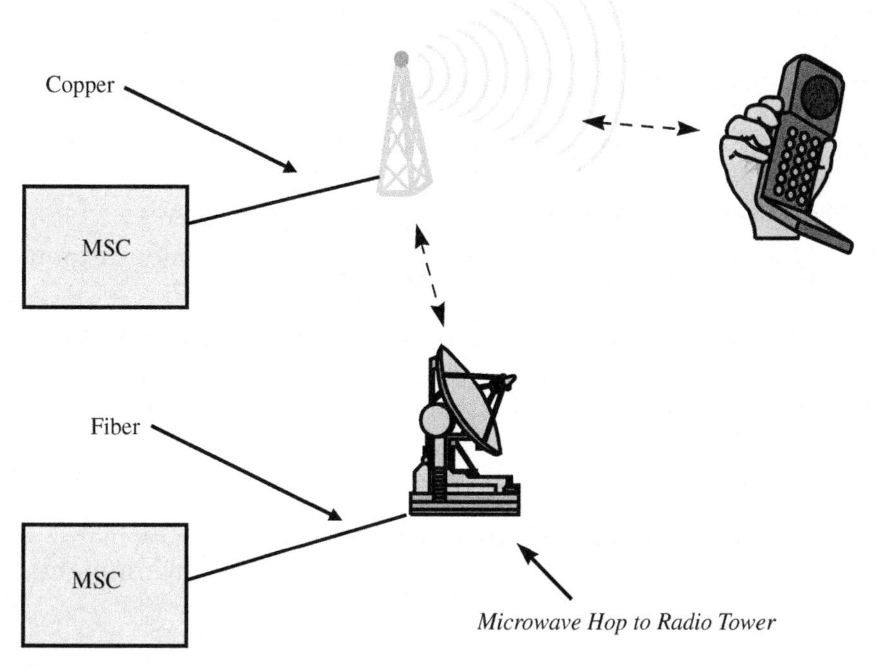

FIGURE 1.5 PCS carrier network.

- The competitors included both new and well established companies.
- PCS carriers must install the latest technology. Cellular carriers must do the same thing, but have their established equipment base.
- PCS carriers had only a fraction of the time needed to make their networks commercially operational. They needed to work hard and fast to capture market share. The PCS carriers had the disadvantage of building a new network infrastructure as rapidly as possible in order to make money as quickly as possible.
- Many PCS carriers had to spend a lot of money to obtain their licenses.

FROM THE PERSPECTIVE OF THE CELLULAR CARRIERS. Unlike emerging PCS companies, the cellular carriers have an existing customer base. However, the problem with being the "big dog on the block" or "king of the mountain" is that there are people nipping at your shoes.

- Cellular carriers have the burden of existing infrastructure to deal with, while PCS carriers deploy new infrastructure—brand new "stuff."
- Cellular carriers must upgrade existing equipment to compete with new carriers coming into the marketplace.
- PCS carriers have the burden of generating enormous expenditures in order to create a profit-making network.
- Cellular carriers must upgrade aging equipment while deploying new infrastructure to meet the needs of an expanding marketplace and new coverage areas.

The major differences between cellular and PCS have more to do with business and less with technology. Figure 1.6 is a representation of the cellular and PCS business environments.

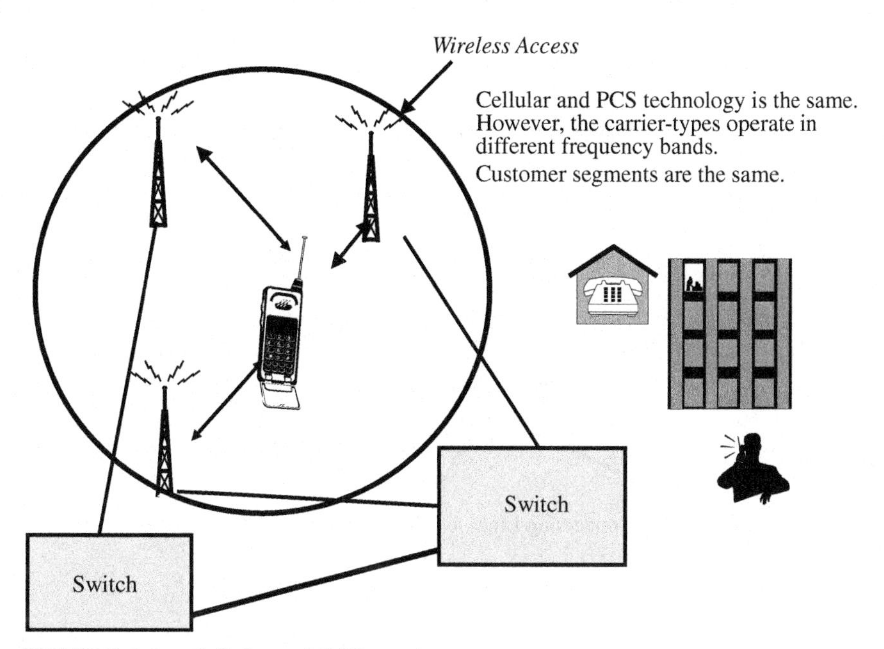

FIGURE 1.6 Cellular and PCS carriers.

PAGING SYSTEMS

Paging is usually considered to be the "low end" of mobile communications. It is less expensive than other mobile communications systems because it is a one-way system. The paging receiver alerts the user to the call but does not verify or respond in any way to the base station. The cost and bulk of a typical mobile transceiver is associated with the transmit portion; this is missing from a paging receiver, which can therefore be small and cheap.

Compared to the cellular and PCS networks, paging networks need far fewer transmitters. The devices are primarily one-way communications devices, although two-way communication is now the dominant form of paging communication. Currently, pagers have limited voice and one-way textual data capability. Figure 1.7 is an illustration of a paging carrier network.

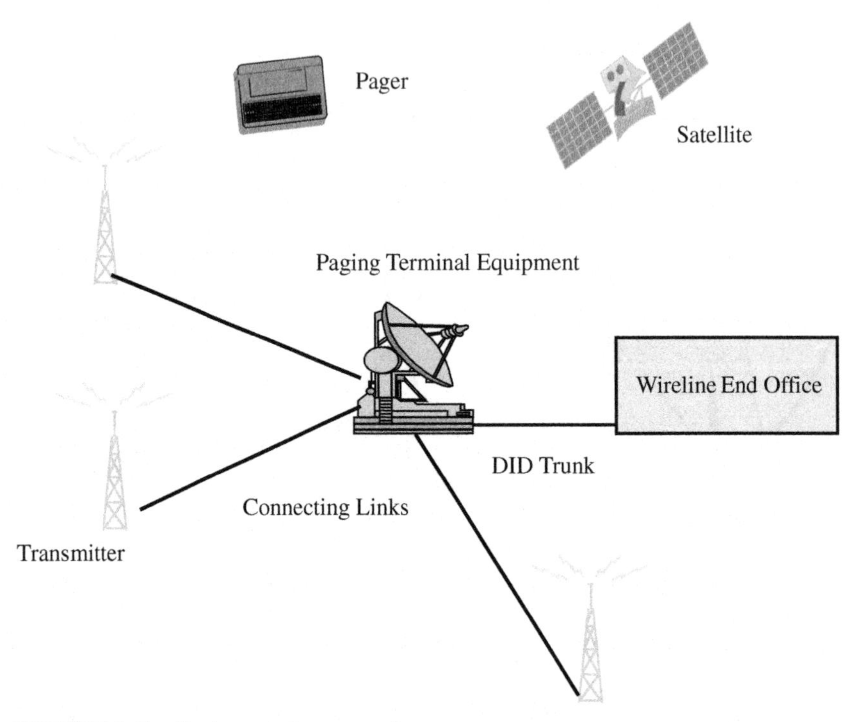

FIGURE 1.7 Paging carrier network.

SATELLITES

Satellite communications is the result of work in the radio field with the objective of achieving the greatest coverage and capacity at the lowest cost. Satellite communications systems can be broken down into two parts:

1. *Space portion.* This includes the satellite and the means necessary for launching it.
2. *Earth portion.* This includes earth transmission and receiving stations.

Unlike earth-based communications systems, satellite communications systems require support in a number of non-communication related areas such as rocket launchers, power supply in outer space, orbital propulsion motors, and more. Figure 1.8 is an illustration of a satellite communications network.

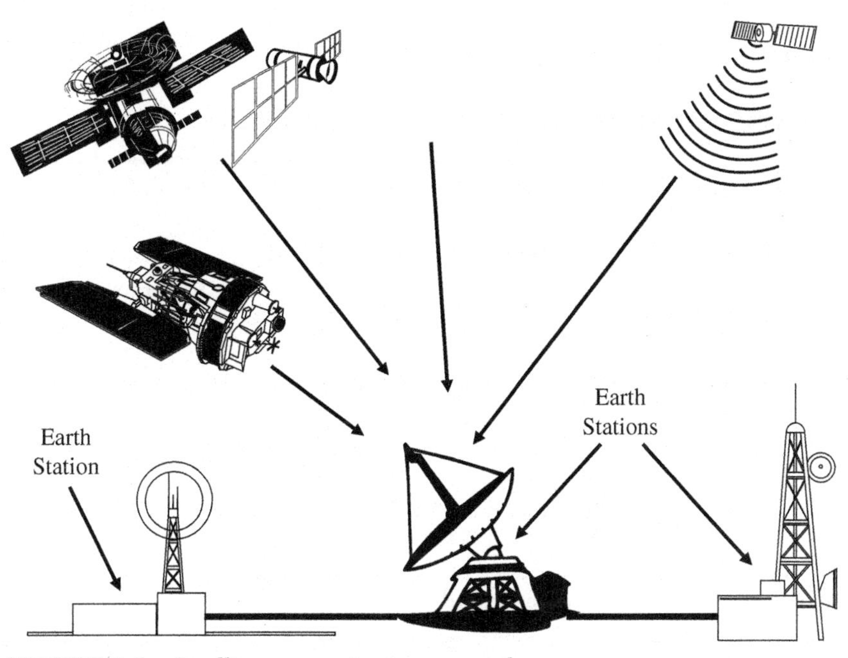

FIGURE 1.8 Satellite communications network.

CABLE TELEVISION NETWORKS

Cable television is one of the world's most popular media and carries many different entertainment channels and clear pictures. The CATV acronym actually stands for Community Access Television, but CATV has become associated with the term cable TV.

Cable television networks are currently designed to transmit multiple conventional analog television signals to multiple subscriber locations. This is a one-way system for distributing the same set of signals to each subscriber location. Historically these systems had a limited capability for return transmissions from designated subscriber locations. The advent of new types of equipment that allow a television to act as both a television and a computer terminal will eventually convert the cable television network into a broadband multimedia information network and connect it to the Internet. Figure 1.9 is an illustration of a cable television network.

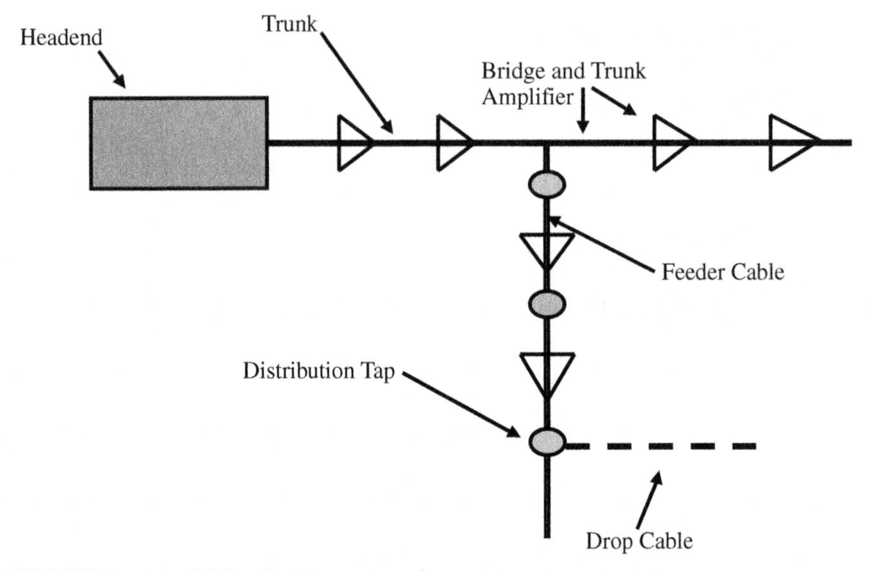

FIGURE 1.9 Cable television network.

Cable television will eventually be capable of supporting two-way voice and video. Whether or not it takes off as a viable telecommunications business will be determined by the marketplace.

INTERNET

The Internet is not a single network but rather a web (network) of networks. Intelligence does not reside within a single component of the Internet but is embedded within the components of individual network elements. When you compare the Internet with the older and traditional voice networks, you will find that the Internet is decentralized in its intelligence. The power of the Internet is in the software supporting the applications and in the protocol itself.

The difference between the Internet and the other networks that have been briefly described is that the Internet is not controlled by any single company. It is extremely popular and heavily used by millions of individuals, and carries voice, video, and data. Today it exists as an overlay network. Physically, the Internet needs the existing wire-line transmission network (primarily under the control of the ILECs and Interexchange Carriers) to carry its information packets. However, the underlying switching/routing matrix is a separate set of network elements typically owned and operated by Internet Service Providers (ISPs). To learn more about the Internet, read my book, *M-Commerce Crash Course*. The Internet has become a catalyst for convergence. Figure 1.10 is an illustration of the Internet.

PROVIDING SERVICE TO USERS (SELLING RETAIL TELECOMMUNICATIONS SERVICE)

The carrier types we have discussed sell service to the individual user or corporate customer. Most consumers are only familiar with those telecom companies that sell services to users and corporate customers. This type of telecom provisioning can be likened to retailing goods.

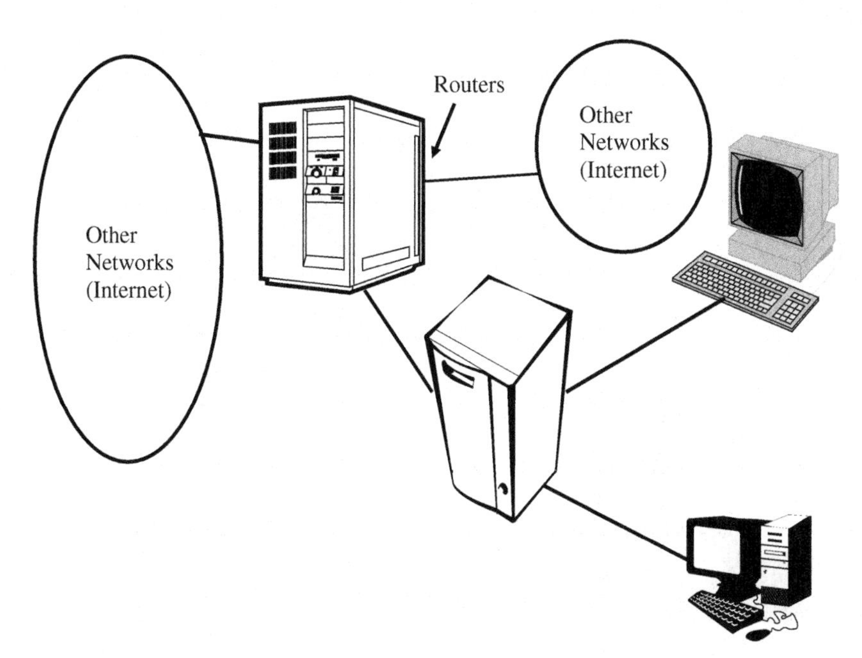

FIGURE 1.10 The Internet.

The transaction between the user and the telecommunications provider is no different from any other retail transaction. Figure 1.11 is an illustration of this relationship.

However, unlike many retail businesses, the telecommunications business is dependent on technology. The technology deployed is not relevant to the user; however, it is the technology that provides the carrier with the product differentiation that the user does see. Technology in telecommunications is changing so rapidly that a company that is not well funded will not survive the competitive wars.

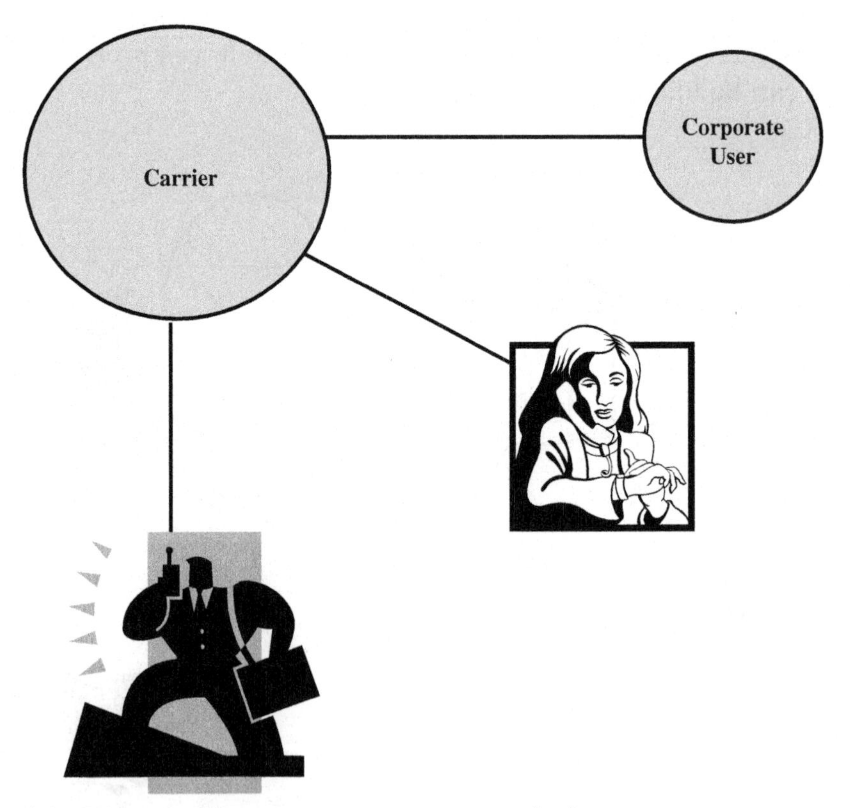

FIGURE 1.11 Telecommunications service: retail sales.

PROVIDERS OF SERVICE TO OTHER PROVIDERS OF SERVICE

Companies that service other companies in the telecommunications business can provide software, labor, calling features support, data storage, and even building services. As a result of recent deregulation in the United States and other countries, the telecommunications industry has given birth to a whole new, rapidly emerging telecom services business. The concept is similar to that of businesses that support other industries. Examples of other support businesses are the restaurant kitchen appliance business, the office supply furniture business, temporary labor companies, data display manufacturing companies, chip manufacturing companies, and the robotic industry. Another way of looking at this is through the concept of the wholesale supply business, where companies supply other companies with goods, but at substantially lower prices than the user could ever obtain. Restaurants purchase a great deal of food; however, the purchases are in such large and frequent quantities that an entire food industry has been created to serve only restaurants.

One can apply this concept to any manufacturing industry or technology industry. In the case of telecom, there are companies that provide specific telecom service support and others that provide real estate space. The following is a list of the types of services provided to telecommunications service providers by other service providers.

- *Infrastructure support.* This includes building space for the switch or network operations center, electricity, water, and HVAC. This is now called *switch hoteling*. Other companies own towers and lease the space to wireless carriers. Many landlords now routinely lease rooftop space for antenna placement. The concept of hoteling was being pursued by the Baby Bells back in the 1980s as way of using a great deal of the excess real estate in their possession at the time. The Baby Bells now call this colocation.
- *Service bureaus.* A service bureau provides any specific service that used to be handled by the telecommunications carrier

itself. Due to rising operating costs, a carrier may see the services of an external company. Service bureaus are now used to support customer care, network management, directory assistance, operator services, or field repair.

- *Switch-based/network-based services.* Companies are now leasing to service providers their own capabilities to create services for the customer. Such a company has its own switch or platform upon which it creates services.
- *Leasing transmission facilities.* As I noted in *Telecommunications Internetworking*, any company that owns its internal communications network can lease that network's excess capacity.
- *Leasing radio spectrum and the associated network services.* Some carriers wholesale their spectrum and their network services.
- *External consultants.* Often, a telecommunications carrier will purchase the services of an outside consultant to help with issues like network design, marketing program development, customer billing, and customer focus group management.

Carriers that sell services or even infrastructure to other carriers are considered "wholesalers" of telecommunications services. Figure 1.12 illustrates various service provider services.

PERSPECTIVE: SUPPLIER AND CUSTOMER

As in any business providing a service, a company can be customer and supplier simultaneously. Think of the food chain. In our food chain, humans ingest animal (this includes anything that walks, flies, swims, or crawls) and vegetable matter. When you review the list of various animal products consumed by humans, consider what those animals eat to survive. Animals eat both one another and plant matter for survival. The telecom business is like any other business; there are those who buy and those who sell. Those who sell want the money of those who buy. This process of selling and buying is also popu-

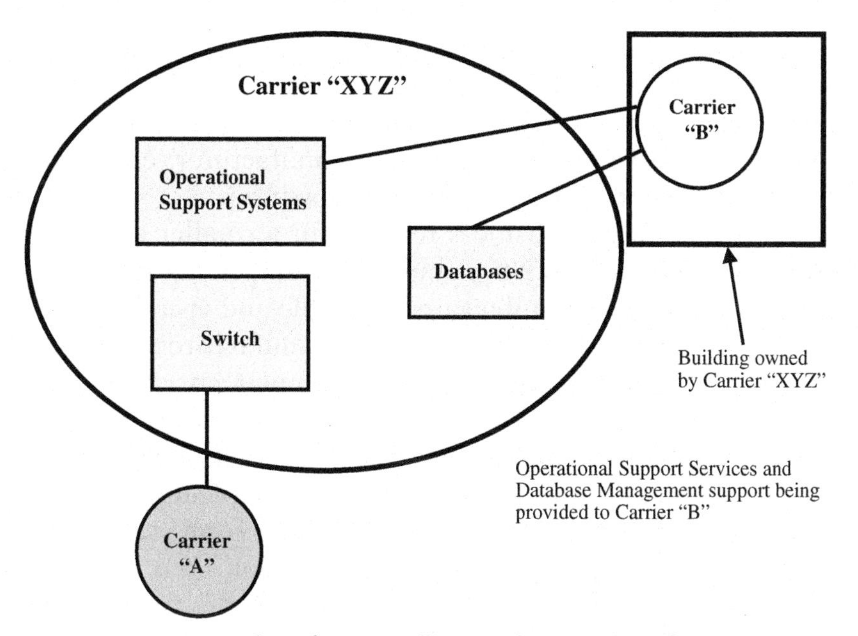

FIGURE 1.12 Providers of service selling to other providers of service.

larly known as the process of *supply and demand*. The customer is at the top of the food chain.

In the world of telecommunications, there are suppliers of equipment and suppliers of services. A sample list of the suppliers of equipment includes the following:

- Switch manufacturers
- Terminal device manufacturers
- Tool manufacturers
- Cellular/PCS base station equipment manufacturers
- Power supply manufacturers
- Battery manufacturers

Typically, service providers (the carriers) like to deal with as few suppliers as possible. There are so many different pieces of equipment that need to be purchased that no single company, no matter how large, could manage numerous business relationships. In order to create an appropriate vendor management environment, manufacturers began to work with one another. The concept of having a single vendor resell another's equipment enables a small manufacturer to have its product sold by a larger manufacturer. The larger manufacturer essentially serves as an agent for the smaller manufacturer.

There are various business reasons for a smaller manufacturer to enter into a relationship with a larger manufacturer and have the larger manufacturer assemble and even install its equipment. Sometimes two or more manufacturers discover that their respective technologies can complement each other, thereby creating monetary value for all. The relationship is called an OEM (Original Equipment Manufacturer) relationship. The term OEM has a broad meaning. Typically, an OEM is a manufacturer that assembles its product from start to finish. The OEM takes raw material, like silicon, and creates a finished product, like a computer chip. An OEM can even take components, like chips, and create a computer. This kind of OEM might be called a *value-added manufacturer*, but one can make the case for its being an OEM because the chip has no value unless it is in a product that people will spend money on, like a computer. In other words, the company is an OEM of computers. The following represents a list of reasons why any company would enter into an OEM relationship:

- The relationship is a strategic one. The smaller manufacturer makes an excellent product and the larger manufacturer needs that product. By joining forces, both can make a sale where neither could singly. The value to the customer is in the combined product.
- The larger manufacturer could be creating a list of suppliers for a particular type of technology and signing up every small manufacturer of that technology, so that the larger manufacturer can say to the customer: "Yes I have relationships with manufacturers of that kind of database. In fact, we have already conducted interoperability testing and even field-tested their product in concert with our switching platform. We have been satisfied with the results." In reality, the larger manufacturer has conducted the same tests with all the other

companies and is satisfied with every single company. Remember, the larger company wants to make a sale and therefore tells the customer what he wants to hear. The smaller company could be one among a list of many companies engaged with the larger company.

- The smaller manufacturer does not have the cash to create and sustain a highly skilled sales and marketing force. Teaming with a larger manufacturer makes good sense, because the larger manufacturer will use its resources to sell the smaller company's equipment.
- The customer prefers to deal with a select number of vendors and finds that OEM relationships create an environment in which customer management of the vendor(s) becomes a driver for "closing of sale."
- The customer has specifically requested that the relationship be established for the value in the companies' working together. Even if the manufacturers dislike one another, they both want to make the sale, so they listen to their common customer.

Those companies that provide voice and/or data service to the user, which enables the process of customer communications with one another, are called carriers and often are also called service providers. Those who work within the telecommunications community tend to look at carriers as being at the top of the food chain because carriers buy all the equipment and services from consultants. A carrier is the entity that has direct and regular contact with the end-user. Vendors of equipment, software, and consulting services typically do not have contact with the end-user. Some software vendors sell in the commercial and private computer markets, which have the kind of widespread presence the carriers enjoy, but those software companies are very few in number. Figure 1.13 depicts the carrier–vendor food chain.

The relationship between vendor and carrier is an unusual one. The vendor believes that it knows the technology better than the customer. The carrier typically responds with a "who

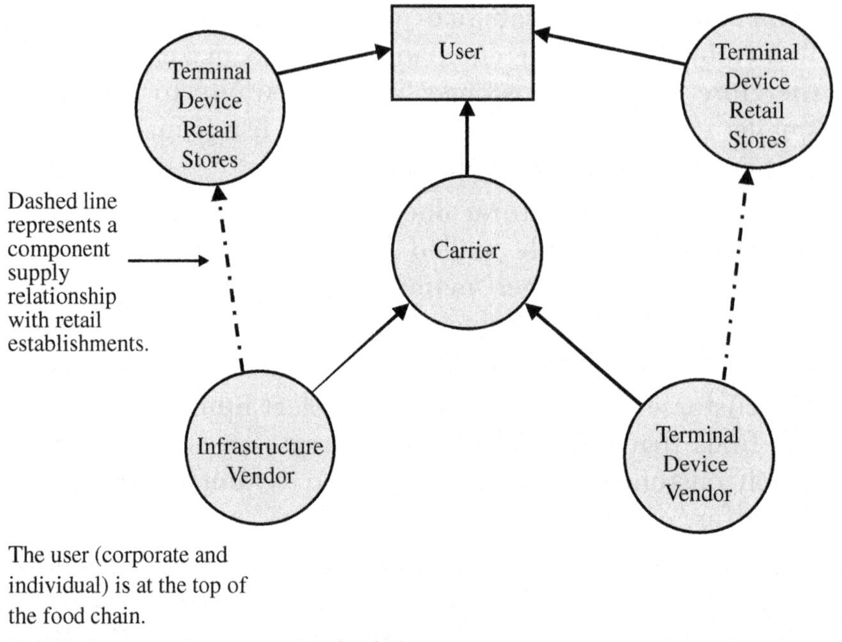

FIGURE 1.13 Carrier-vendor food chain.

cares?" attitude. It is a fascinating study in one-upmanship. The carrier believes it wins while the vendor believes either that it has pulled the wool over the carrier's eyes or that the carrier has beaten it into the ground. Carriers and vendors play a constant game of winning and losing. To a carrier, the concept of a win-win relationship is very much a reality. To some vendors, there can never be a true win-win relationship. The vendor sells boxes or software (or license fees) at a certain price to maintain a certain margin. The carrier always wants to pay less and get more value than the vendor feels comfortable with. The relationship between a carrier and vendor can run the gamut of feelings that any two friends may have toward one another. Some carriers have been known to run vendors through exhaustive analyses, the result of which is either a purchase or no purchase at all.

In the end, both realize they need each other and neither can survive unless the other is healthy. To that end, both carrier and vendor realize that the end-user must be satisfied. The end-user (customer) is ultimately the only part of the food chain that counts. Figure 1.14 is an illustration of the end-user's role in the telecommunications revenue cycle. The process of identifying user (customer) requirements is part of the product and sales management process, which we will discuss later in this book.

Selling equipment or software to a carrier can be very rewarding to those sales people who love to close deals, even unprofitable ones. To others, it is an exhaustive process requiring knowledge of the customers of the carrier and the personalities in the carrier. Service providers will compare vendors of products and services, seeking both current and future value in the product and in the vendor relationship.

WHY IS THE CARRIER–VENDOR RELATIONSHIP IMPORTANT?

The relationship between a carrier and a vendor (of products and services) can and should be both strategic and tactical. A

FIGURE 1.14 The end-user.

close relationship, in which the parties communicate freely with one another, can enable a carrier to react quickly to rapid market changes or even service crises. Sometimes a carrier is forced to do business with a vendor simply because the vendor is the only supplier of the product. However, in such a situation, the carrier usually begins a long process of finding a replacement product or technology.

The benefits of a close carrier–vendor relationship are:

- The carrier can react quickly to market changes. Vendors who like working with customers will react more quickly to a carrier's last-minute needs. It is human nature for parties to establish a level of communication, which is close and frequent, when the parties like one another. When the communication level between the parties is so close, the vendor will often have an inkling of what new work or marketing initiatives are being planned. Of course, the vendor's closeness is the result of keeping the customer satisfied and helping the carrier meet its revenue goals.
- The vendor in the established relationship will often get the bulk of the business, which is good for the vendor.
- The carrier will often have early access to new technology or products when it has a long-term relationship with a vendor.
- The vendor will get a lot of business from a carrier if it provides its largest customer with early access to technology and products.

Note that the above list represents the "give and take" of the carrier–vendor relationship. The real importance of the carrier–vendor relationship is that each needs the other. Figure 1.15 is an illustration of the carrier–vendor relationship's value.

INTERDEPENDENCY

Telecommunications companies provide equipment, software, and service. Their customers are those that supply equipment,

The relationship between a carrier and vendor is key to the vendor's financial success.

- A big carrier sale is news that enables a vendor to sell its product to other carriers.
- Carrier can be used as a reference.
- A large enough carrier gives a vendor instant media attention.
- Vendors need the carriers.

FIGURE 1.15 Carrier–vendor relationship value.

software, and service to others. The telecom industry is comprised of companies that are both suppliers and customers. The same can be said of any retail or wholesale industry; someone is seeking services or products, while someone else wants to provide services or products.

Unlike many other industries, the telecom industry is one that is highly dependent on technology. Given this dependency on technology, the telecom industry is a dynamic and interactive community, where the exchange of ideas is routine. The ideas exchanged include marketing practices, regulatory information, technical data, and even business models. Telecommunications has been a part of people's lives for over a century. In order to ensure that customers around the world can communicate with one another, national and international standards have been established. In reality, every nation uses different network signaling standards. However, the industry's heart is in the right place, because countries all use common conversion standards so that their inhabitants can communicate across national borders. Telecommunications is a part of all our lives and there is an interdependency among the players in telecommunications.

Each company is providing a service. Each company plays more than one role in the provisioning of telecom services.

Ultimately, the end-user or corporate customer is the beneficiary of the telecommunications companies' work. Figure 1.16 is a depiction of the industry's interdependency.

THE CULTURE

Telecommunications companies include those that work in wireline, wireless, Internet, web hosting, satellite, cable television, paging, broadcasting, and even the old citizens band radio. Telecommunications is about "communicating" between people. The technical and business fundamentals behind operating a telecommunications company are all the same; no matter what the telecommunications industry segment is.

Understanding the relationship between the supplier and service provider is important to understanding how the indus-

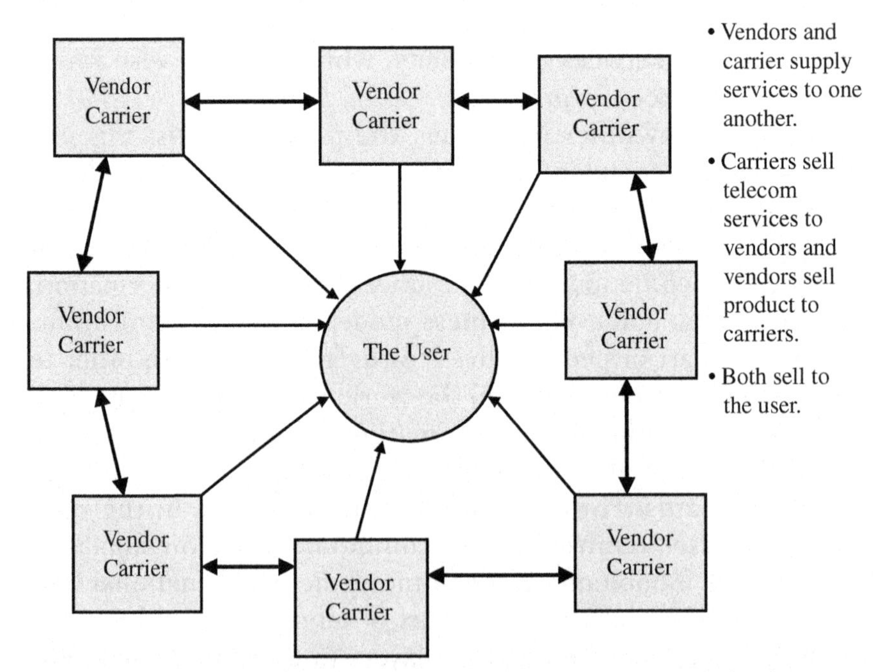

FIGURE 1.16 Companies are both suppliers and customers to each other.

try functions. The relationships I have described are part of the way the industry operates. Many outside the industry may find the intertwined relationships of simultaneous competitors and partners difficult to comprehend. We should remember that this business started intertwined. At the turn of the twentieth century, there was one Bell System in the United States; at one point even Canada was part of it. Globally, every nation had its own telephone company and telecommunications manufacturer. This is called a *monopoly* business environment.

The Bell System was a big and argumentative family, brought up to discuss all issues pertaining to providing service to the customer. The key word or operative word for the members of this family was "service." The people brought up in the business during those days were taught that providing service was the only priority. This philosophy had even permeated the operation of telecom manufacturers. Providing telecom service was a part of life.

"Service first" was the culture of the AT&T Bell System and every other telephone company in every other country. However, since the 1984 divestiture of the AT&T Bell System in the United States, the elements below have disappeared from the telecom community:

- Common corporate culture
- Common training
- Common practices
- Common goals

The common goal was service, not revenue generation. Revenue generation was guaranteed due to the fixed rate of return guaranteed by the state utility commission. The telephone companies before 1984 and before the Telecom Act of 1996 were measured by clearly defined service objectives/parameters. One old service measurement was "dial-tone delay" (the time it took a customer to hear a dial tone from the moment the telephone was lifted off of its cradle). The dial tone was an indication to the customer that service was available. The reality of providing service at any cost was a fact of life.

Today, providing service at any cost is unheard of. Today service must be provided within defined financial parameters. The Telecom Act of 1996 changed the landscape of the telecommunications business. Telecom companies no longer have:

- *Common corporate culture.* People used to start careers in the factory making the equipment, in operations, or in a general department. Today we have specialists in one area and no broad training (with depth in one or more areas).
- *Common training.* We all used to go to the same school in Illinois.
- *Common practices.* The practices were called the Bell System Practices (BSPs) and there used to be a journal called the *Bell System Technical Journal.*
- *Common goal.* Telecom companies compete against one another.

TELECOMMUNICATIONS PROFESSIONAL AND TELECOMMUNICATIONS MANAGER

No matter what industry segment or company type we address, two new types of telecom workers have emerged: the telecommunications professional and the telecommunications manger.

The telecommunications professional is a generic name for the telecom jack-of-all-trades. The term is fairly nondescript, but in today's employment market, survival is key. A variety of experience is no longer considered a negative. The telecommunications manager is a professional who is a manager with a broad set of management skills and a broad understanding of the business.

SUMMARY

Competition is good for the nation. The challenge for the telecommunications companies is to compete without sacrificing customer service.

Prior to the Telecom Act of 1996, people worked for the telephone company, the long distance telephone company, a wireless carrier, or a paging company. A telecom person was identified by the segment of the industry in which he or she worked. Today, one is called a telecommunications professional. Telecommunications management is now taught in graduate schools as part of part-time weekend programs. The world has changed, and with it the environment in which the telecommunications professional functions. What is evolving, out of the necessity of career survival, is the telecommunications manager who has a broad understanding of the telecom business. These professionals are forced to obtain these skills via new job experiences.

This book will look at the relationship of the carrier with the vendor, the internal decision-making processes of both and how decisions affect both, and management processes. Chapter 1 established the players in the business. Chapter 2 will examine the organizational structures and answer the question: What kinds of organizational structures are used in a telecommunications company? We will find there are sets of basic organizational structures that can be used by any company. The structure used by a corporation is dependent on the type of product or service the telecom company is providing. How well the organization functions is dependent on how good the management team is, both individually and as a cohesive management group. We will touch on management styles throughout the book. One cannot talk about organizational structures unless one understands management practices and styles.

CHAPTER TWO

ORGANIZATIONAL STRUCTURES

Some years ago, an organizational theorist told me: "Organizational structure is not just a picture of boxes; rather it is a pattern of interactions and coordination that links tasks, technology, and the human components of an organization together to ensure that the organization accomplishes its purpose."

CLASSIC ORGANIZATIONAL STRUCTURE

Over the years, a great deal of work has been done to try to understand how the person and the organization interact. The classic organizational structure is typically characterized as highly bureaucratic. Most people view the government or large corporations as bureaucracies. Experts on organizational behavior have developed theories on how a bureaucracy operates. A model promoted by some organizational theorists states that a bureaucracy is characterized as a management structure that has the following major characteristics:

- *Specialization and division of labor.* Max Weber, a pioneer in modern sociology, stated that a bureaucracy contains, "A specified sphere of competence. This involves a) a sphere of obligations to perform functions that have been marked off as part of a systematic division of labor, b) the provision of the incumbent with the necessary authority, and c) that the necessary means of compulsion are clearly defined and their use is subject to definite conditions."* What Weber was talking about is that a company or government agency divides work into specific lines of authority and specialization in order to carry out its mission.

- *Positions are arranged in a hierarchy.* The hierarchy of management needs to be discussed briefly. An organization of any size should have some kind of "pecking order." Someone must be held responsible and accountable. The concept of responsibility and accountability will be discussed later. The concept of a management hierarchy follows the principle that each lower office is under the control and supervision of a higher office.
- *System of rules.* Most people who work for a living, including experts on organizational behavior, believe that an organization of functions (a company) requires rules by which it must be bound. Rules are needed to establish an orderly flow of communication and the function of processes. The official rules facilitate the creation of a corporate culture comprised of both official and unofficial rules. The unofficial rules in a company are, in fact, social rules that guide how employees communicate with one another, how management communicates, and how the people in the company view themselves and the company in the world.
- *Impersonal relationships.* The relationship between management and the employee was once considered an impersonal one free of emotion. Communication between management and the employees was limited. Today, management experts stress compassion and respect for the employee. However, the relationship is still impersonal because it cannot cross an imaginary line drawn by the abstract concept of professionalism.

*A.M. Henderson and Talcott Parsons, "Max Weber: The Theory of Social Economic Organization," Free Press, NY, 1947, p. 330–340.

The reality is that the characteristics just noted exist in any company to varying degrees. These characteristics were a popular view of organizational structures until the 1970s. Many may feel this view still exists. Remember that there is theory and there is reality. An organizations is made of people, and the rules by which it functions result from the current social environment and the industry the organization works in. Figure 2.1 is an illustration of the classic organizational structure.

I addressed the classic view of organizational structures in order to establish a baseline of understanding for the reader. Theorists believed that an organization should be completely free of emotion. However, that has never been the operating reality, even when the theory was popular. Emotion has always been a part of the workplace. Lack of compassion for employees is an emotion called indifference and coldness. Before the 1970s, the labor unions were so powerful that a labor strike was the primary method of communication. Social and economic needs required change, and companies now try for a kind of friendly and supportive environment, which helps to prevent labor disputes and is treated as a positive attribute of the company when employees are hired. The difference today between the classical theory of organizational management practice is compassion and respect. Figure 2.2 is an illustration highlighting the emotional aspect of today's company structure.

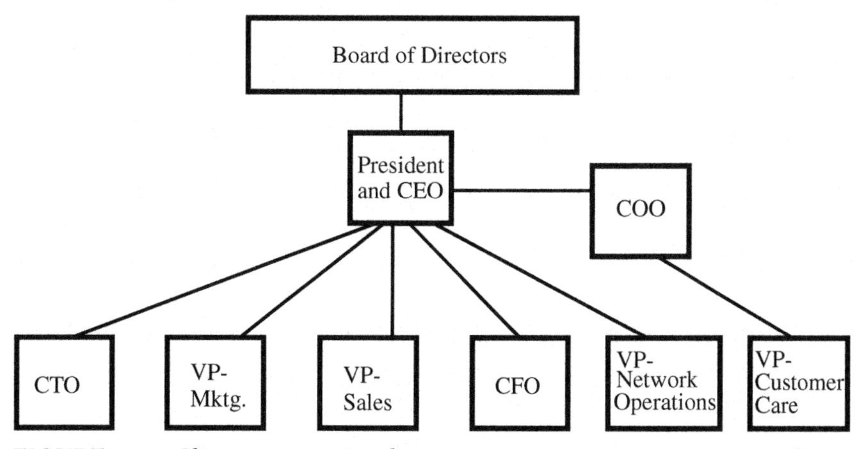

FIGURE 2.1 Classic organizational structure.

- Employees are important
- Teaming
- Training
- Openness
- Employee suggestions are important

FIGURE 2.2 Compassion and respect in the workplace.

Organizational theory is a complex topic and requires its own book. However, organizational theory needs to be considered in this book in order to understand the practical realities of this theory and its application to the telecom world.

DECISION-CONTROL PROCESS AND ORGANIZATIONAL STRUCTURE

I believe the most significant organizational behavioral change is the way in which companies manage the decision-making process. Decisions are made in two ways: in a centralized or decentralized fashion. The *centralized* decision-making process has been in existence for centuries and was the form of decision control used in the business world for many decades. *Decentralized* decision making requires senior management to trust less-senior management. The decision-control process is the core of how a company operates and is a strong indicator of how management and employees relate to each other (corporate culture).

This book will continue to delve into organizational and management theory where appropriate.

As I noted, the decision-control process is a reflection of the company's personality and involves such issues as:

- The management–employee relationship
- Management policies
- Human resources policies
- The benefits plan
- Organizational structure
- Public relations

We will understand how the decision-control process affects the company as we read through this chapter, because the process permeates the very fabric of a company. Figure 2.3 is an illustration of the centralized and decentralized decision-making processes.

CENTRALIZED DECISION MAKING

Typically, a structure is applied across a company and not just in one organization (department or group). To have multiple structures applied in a company would cause anarchy. Running a company is difficult enough considering the multitude of personality types working in the company; applying multiple management structures to the overall corporate framework would be disastrous.

Centralized decision making was common many years ago. It became unpopular as corporations grew to national and international size (see disadvantages of centralized decision-making, below), but it is making a comeback in the telecommunications industry. Centralized and decentralized decision making each have advantages and disadvantages. The advantages of centralized decision making are:

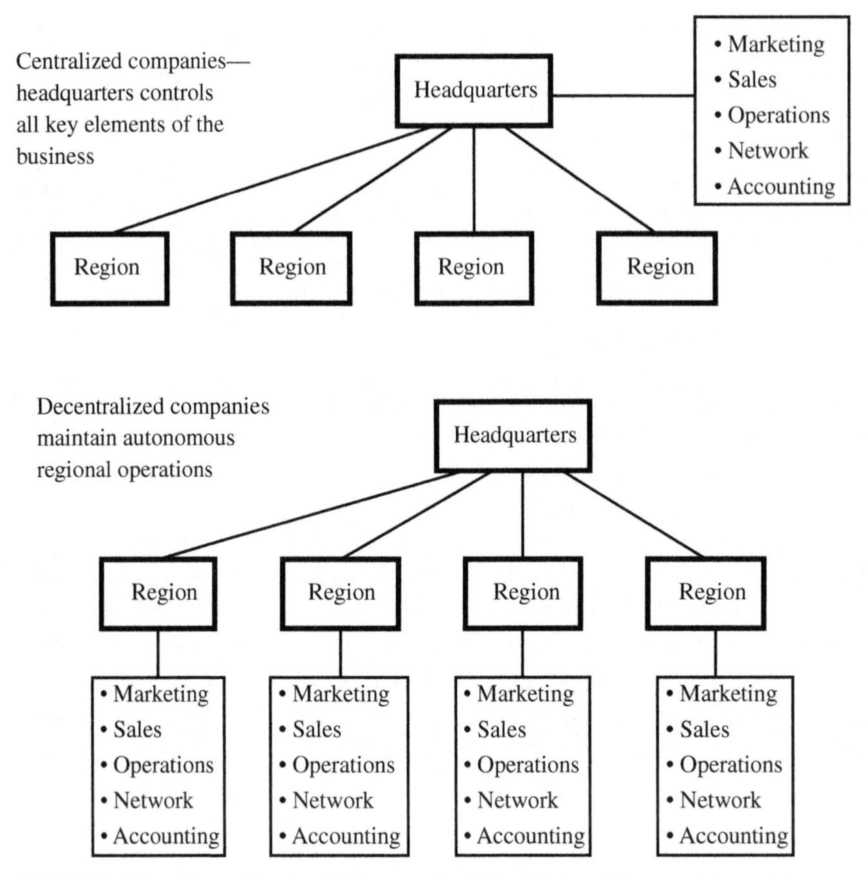

FIGURE 2.3 Centralized and decentralized decision making.

- A single set of orders is issued
- There is a single set of standards: operational, financial, human resources, and management
- Rapid decision making is possible
- Management is focused

The disadvantages of centralized decision making are that it:

- Can stifle creativity
- Can easily lead to "kingdom creation"

- Can lead to poor communications because of the chain of command
- Can lead to slow decision making because of fear reprisal from upper management

Centralized decision making tends to lead to vertical organizations with multiple levels of management, even though instructions are issued from the "top." However, there are "flat organizations" with few levels of management and just a handful of decision makers. In a flat organization, the instructions are still issued from a single person (i.e., the top of the company). Figure 2.4 is a representation of the advantages and disadvantages of centralized decision making.

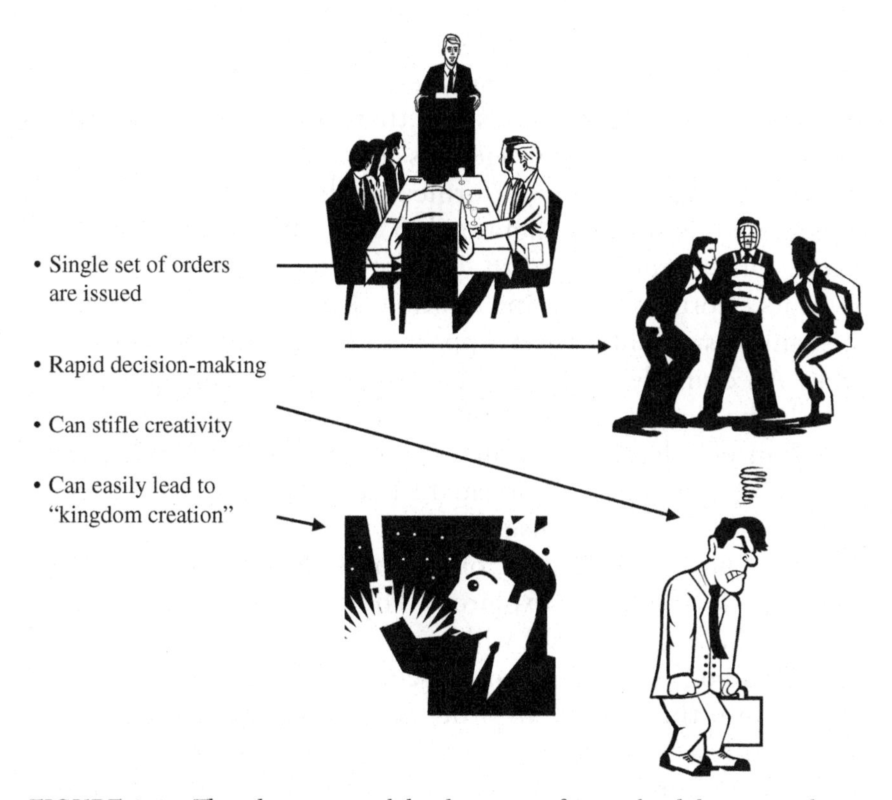

FIGURE 2.4 The advantages and disadvantages of centralized decision making.

ADVANTAGES OF CENTRALIZED DECISION MAKING. The benefit of a centrally based company is that there is only one "king" or "queen" in charge. For those who have never managed organizations in real life, remember this: companies are not democracies. Running a company by committee is an ingredient for that recipe called "disaster."

As I had noted, the advantages of centralized decision-making are:

- Single set of orders or instructions are issued
- Single set of standards: operational, financial, human resources, and management
- Rapid decision-making
- Focuses management

SINGLE SET OF INSTRUCTIONS. Schools teach the science of management. However, managing is not just science, it is also art. People have personalities and their own personal agendas, and often these personal agendas are hidden. In a company, people have tasks to be performed at their assigned time. Of course, no one wants a dictatorship, but without order there is chaos. One cannot afford the luxury of having a company defocus its efforts by executing multiple sets of instructions about the same issue when these appear to be in conflict with each other. An example of this is when the president of the company tells the vice president of business development and the vice president of sales to determine if a particular market segment should be targeted. The president knows that both vice presidents are in complete disagreement over the proposed target market. The vice president of business development believes the market should be explored. The vice president of sales believes the target market will be a money loser. The president has asked both vice presidents to conduct independent investigations into the matter. In order to conduct proper investigations, both will need to work with the vice presidents of technology, marketing, and finance. The investigations cannot be conducted in isolation.

Operating a telecommunications company is an exercise in departmental integration, but imagine the president doing this all the time. Having the occasional disagreement is fine. However, having decisions made in this manner all the time is an ineffective way to run a company. The vice president of business development is in charge of investigating new opportunities and should be working with the vice president of marketing. The vice president of sales may have personal connections in the targeted market segment and should be consulted. Working together is important; however, the president should not have placed both vice presidents in charge. It is true that the occasional leadership teaming effort sometimes enables the management team to bond. However, when it is done as a matter of normal business, there will be problems. Imagine that the example I described occurs all the time, with the various vice presidents' initially disagreeing with one another. Working together will not necessarily change their opinions on the issues. After a while, this management style encourages the vice presidents to undermine each other. (Remember that one of the ways vice presidents are assessed is by the number of wins with which they can be credited.)

The best way to understand the impact of a single set of instructions for a company is by understanding how multiple sets of instructions can adversely affect a company. Even if a company had enough employees to act on multiple sets of orders, the following disagreeable things could occur:

- *Unfavorable public perceptions.* Confusion could reign among customers about what the company is either selling or executing.
- *Strained company resources.* Companies attempting to act on multiple sets of instructions could find internal support organization resources, like technology development and marketing, strained beyond limits.
- *Poor execution leading to failure to meet goals and objectives.* The likelihood is high that the company's forces would execute the multiple sets of instructions poorly. The company's employees are not incompetent; the reality is that people can only multitask to a point. Executing instructions to support

internal activities or external activities requires a great deal of coordination.

- *Confusion among employees.* Employees are key to a company's success. The execution of multiple sets of orders or instructions not only communicates a message to customers but also communicates that same message to employees. Confusion can lead to low morale, low confidence in management, or both. By themselves, low employee morale and low confidence in management are devastating to a company.

Figure 2.5 is an illustration of the process of issuing a single set of instructions.

SINGLE SET OF STANDARDS; OPERATIONAL, FINANCIAL, HUMAN RESOURCES, AND MANAGEMENT. Internal corporate standards are critical to the ongoing mental and physical health of employees. Well-thought-out internal corporate processes create an environment that either enhances or annihilates an employee's ability to perform. Good operational, financial, human resources, and management practices and standards enable a company to produce a product but also creates a working environment that enables employees to do their jobs well.

Operational standards include processes and procedures that focus on the internal and external use of resources. These procedures and processes address the following areas:

- Confusion reduced
- Proper management of resources
- Public perceptions can be better managed
- Company resources will not be strained
- Likelihood of poor execution is reduced

- **A single authority or source of direction is important to the overall health of a company.**
- **Telecommunications carriers are organized like armies.**

FIGURE 2.5 Issuing single set of instructions.

- Overall quality of the product and the performance of the employees
- Enabling and facilitating communication between departments
- Consistency in product quality
- Consistency in interdepartmental communication
- Consistency in intradepartmental communications
- Inventory management
- Product management—Translating customers' needs into technical requirements toward which the factory can build
- Product development—Building the product to the customers' requirements
- Marketing—Identifying the correct market segments, defining the product, and creating the message
- Sales—Creating a consistent message to the customer means making sure you are selling what you are making.

Operational issues are highlighted in Figure 2.6.

Financial operational standards concern the establishment of cost controls and other internal financial mechanisms. These specifically address the following:

- Quality control
- Communications between departments
- Inventory management
- Product management
- Product development
- Sales
- Marketing
- Engineering

A set of performance criteria needs to be maintained for every department area.

FIGURE 2.6 Operational standards.

- Purchase of supplies and other parts
- Travel and entertainment (T&E) expenses for employees, especially the sales force
- Purchasing and T&E approval authorities—Only a handful of managers are entitled to approve certain purchases and T&E expense reports.
- Approval levels—Employees should be able to trust managers to approve expenditures at specific dollar values. This enables management to delegate authority and optimize the management decision-making process within the company.
- Overall cost controls for the expenditure of money—This does not mean that the finance department decides what can or cannot be purchased; rather it means that criteria are put in place to assist managers in deciding how much to buy.
- Approved vendor lists—The approved vendor list is a short list of vendors from whom the company purchases services and equipment (parts). Every company has its own criteria for determining which vendors are placed on the list. Most companies forbid employees to purchase even office supplies from any company not on the approved list.

Figure 2.7 represents the benefits of having common financial standards.

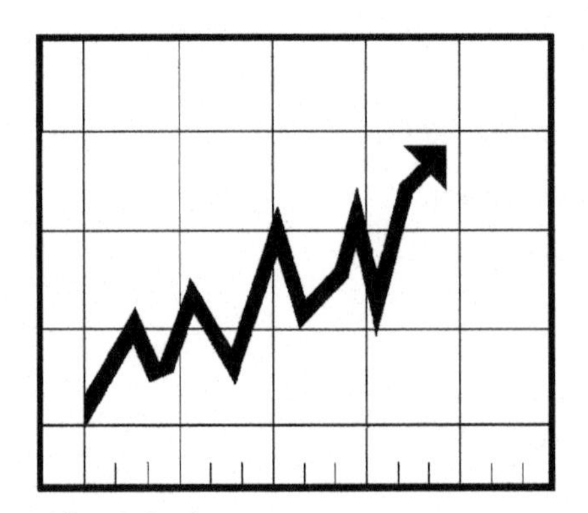

FIGURE 2.7 Common financial standards.

Human resources issues are managed by the company's Human Resources department and can range from benefits to employee safety. Such issues include:

- Health benefits administration and the constant upgrading of employee health benefits
- Life insurance administration for company-provided life insurance policies
- Disability—Long and short-term disability insurance is overseen
- Vacation
- Sick leave
- Payroll taxes
- Reference checks
- Employee aptitude testing
- Training administration—Sometimes a department will administer its own training. However, in most companies training is administered by the Human Resources department to ensure that employees' records are kept up to date.
- Storage of employee personnel files
- Implementation and oversight of the company's code of employee conduct
- Managing interview schedules for prospective employment candidates
- Managing corporate-wide employee communications
- Overseeing the process of manager–employee conflict resolution
- Salary administration—This includes managing all aspects of the employee payroll.
- Ensuring a company's compliance with state or provincial employee and work environment regulations. This includes building code compliance, environmental regulation and law compliance, and overall on-site employee safety. An example is a situation in which a company in Pennsylvania had a roof

leak from rain that caused the carpet to become wet. The wet carpet went untreated and then suddenly a mold condition developed. The company immediately took action to evacuate the specific work area and have the area cleaned. This entire effort was under the jurisdiction of the Human Resources department.

- Managing employee safety

The Human Resources department in all companies has broad range of responsibilities. In some companies, employee safety may be managed by a group called the Employee Safety department. However, this group operates under the jurisdiction of the human resources organization, or the same vice president oversees both Human Resources and the Employee Safety Department. Figure 2.8 is an illustration of the Human Resources department's function.

The job of Human Resources is to tend to the common needs of all employees, including the executive management team. In small companies, corporate security falls under the Human Resources organization. In very large companies, corporate security may be in its own department. In large compa-

FIGURE 2.8 The Human Resources department.

nies, corporate security is a common issue for the Human Resources department and the Corporate Security department.

In general, the employee must find the work environment comfortable and safe. The Human Resources organization works to create an environment that meets the needs of the employee as well as those of the company. Human resources issues are treated the same throughout the company, no matter what its size. If employee issues were treated differently in different corporate departments, the corporation would run an enormous risk of being sued by its employees for a variety of discrimination matters. Figure 2.9 is an illustration of the Human Resources department's overall function.

Management standards refer to how management has decided to interact and communicate with the employee population on all company matters. They are a combination of definitive practices and the "feel" the executive management team wants conveyed throughout the company. These management standards are the "look and feel" of the company as perceived by the employees and the customers. This "look and feel" encompasses all the items in this section and the leadership style of the senior executive team. This is the one area of corporate practices in which everything tangible and intangible about running a corporation converges. Management standards include:

Balancing the Needs of the Company and the Needs of the Employee

- Pensions
- Benefits
- Vacations
- Employee rights
- Safety regulation oversight

FIGURE 2.9 Overall function of Human Resources.

- *Management team professionalism.* Professional attitudes must exist and be seen by the employees and the customers. Professionalism is an area that encompasses management behavior, employee treatment, ability to communicate, ability to execute on actions, consistency, and knowledge.
- *"Walking the talk."* When the executives make a promise, they had better keep it.
- *Openness to suggestions.* If the management team is not open to employee feedback, it should at least seem to be interested. However, pretending to respect employees' opinions will not last for long; employees are not stupid.
- *Respecting the employee.* Respect for the employee must be demonstrated in a public fashion.
- *Acknowledging employees' contributions in a public fashion.*
- *Managing by walking around.* The term "managing by walking around" is an old one that refers to a style of management in which a company's managers are seen and heard on a frequent and regular basis by the general employee population. In my experience, if this does not happen, employees will have a variety of negative feelings, including isolation, lack of respect, lack of information.
- *A single mode of behavior and communication is critical and is easier to achieve in an organization that is centrally controlled.* It is far easier for the president of the company to control a handful of lieutenants' behavior than to control the behavior of a few dozen nearly autonomous regional and departmental managers (as in a decentralized company).

This area of practice in fact is mostly "management art" and less "management science." The principles of management can be taught in school, but leadership and the practice of management require experience. A common face and common set of management principles are far more easily presented and achieved if the company is centralized in its decision-making structure. Figure 2.10 is an illustration of the points made about management principles.

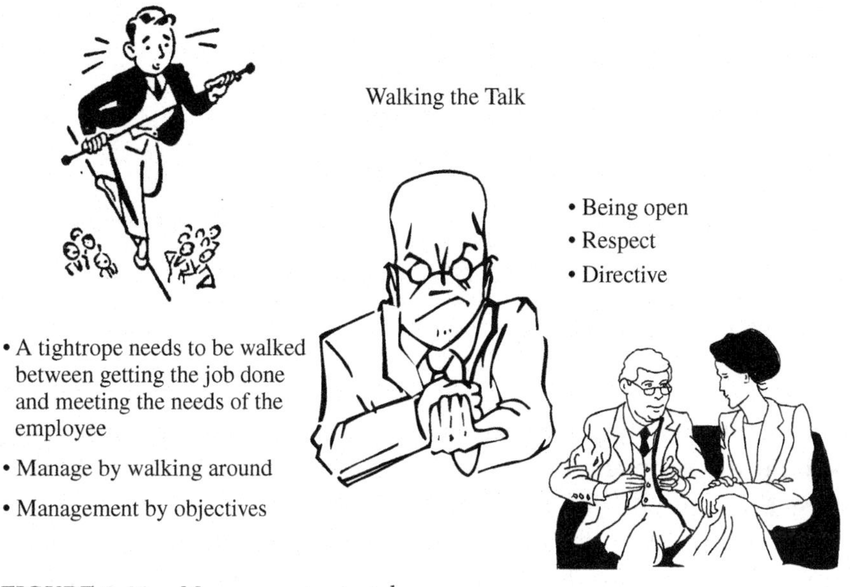

FIGURE 2.10 Management principles.

RAPID DECISION MAKING. Decisions are normally reached far more quickly in a company whose authority is centrally structured. This situation is achieved because of the small number of people needed to obtain support. It may sound dictatorial, but the reality is that companies cannot be run by consensus. Consensus is necessary to gain internal political and operational support for decisions that affect the entire corporation. However, at the end of the day, only the person in charge can make the final decision.

Centralized organizations keep major decisions restricted to a few people, though input is gathered from many. Figure 2.11 is an illustration of rapid decision making due to centralized organizational structures. It does not always work, but we will delve into that later in this section.

FOCUSED MANAGEMENT. A centrally managed organization can optimize resources by focusing specific talents and resources on issues that affect the entire company, including remote locations and regional offices. By having a central place where problems can be solved and issues addressed, a core group of talent can be focused on helping the whole corporation. This is what a

Responding to a Central Authority	• Rapid Decision Making is a result of a centrally managed organization • Fewer people involved in the decision • Consensus is not a required element

FIGURE 2.11 Rapid decision making.

staff group is supposed to do. A central staff that works well should focus on the big issues that affect multiple lines of business throughout the corporation.

A central staff does not need to be large but it needs to work toward solving problems that impede the operation of various organizations within a company or within a single organization in the company. Many people believe that the only purpose of a staff is to manage the "care and feeding" of executives. Unfortunately, far too often executives use their staffs to serve themselves rather than to serve the organization for which the executive is responsible.

An example of a highly functional staff would be an engineering staff working in a telecommunications carrier. The staff is responsible for reviewing new technology, establishing common design practices, working with industry groups to develop technology standards that benefit the industry as well as the company, overseeing trials, overseeing massive systemwide technology deployments, establishing the technology evolution of the network, determining resource allocations, determining promotion candidates, serving as the department's single point of contact with other departments, overseeing the retirement of technology or specific switching centers, training for all departmental personnel, supporting joint departmental efforts, representing the organization in regulatory matters, managing the organizational budget, auditing operational areas, determining what projects will be implemented, and determining staffing requirements. A staff that performs these functions enables the regional offices to focus their energies on meeting specific market segment needs and managing day-to-day operations. Figure 2.12 is an illustration of the ideal engineering staff.

A centrally managed organization also focuses management on their assigned tasks. There is very little chance of regional offices' usurping the authority of headquarters. On many occasions, remote office locations left to their own devices have wreaked havoc on a company's overall operations and public image.

Centralized decision making enables management to focus on critical matters and not clean up after a manager who created a mess in the process of trying to make a "name" for himself. Organizational issues can be major distractions for a management team.

There have been instances in every company that has ever operated, from startup through the mature company, where management has been forced to deal with issues that are nothing more than distractions. These distractions have usually centered around in-fighting for power and control among management groups. The more decentralized the management structure, the greater the likelihood that an ambitious manager will take advantage of the dispersed nature of decision making either to create a crisis whose solution will make him look "heroic," or he will try to solve a problem that headquarters is already addressing and thereby get in the way of progress.

- The staff group oversees special projects
- Reviews technology
- Establishes organizational standar
- Solves problems that impact the entire organization
- Oversees multidepartmental activities

FIGURE 2.12 Example of a staff group.

FIGURE 2.13 Personnel management in a centrally managed company.

Figure 2.13 is an illustration of the personnel role of a centrally managed company.

DISADVANTAGES OF CENTRALIZED DECISION MAKING. For every advantage there is a disadvantage. As I noted earlier, the disadvantages of centralized decision-making are that it:

- Can stifle creativity
- Can easily lead to "kingdom creation"
- Can lead to poor communications because of the chain of command
- Can lead to slow decision making because of fear of higher management reprisal

STIFLING CREATIVITY. A danger to any centrally managed company is the stifling of creativity among employees. Centrally controlled companies, when not properly managed, can either strike fear into their employees or discourage creative input. Centrally managed companies focus activities affecting the corporation in the headquarters arena; employee input in the regional offices or even within the ranks of headquarters is not necessarily sought.

Sometimes input from the "field" is requested. However, headquarters often takes this input but does not act on it, or acts on this advice or input in a way other than what the field had anticipated. Part of the problem in this instance is that headquarters usually does not tell anyone in the field why or how a decision was arrived at. This lack of communication, whether deliberate or not, tends to discourage employees from spending any time providing suggestions. People often want to know why their ideas have been rejected or ignored.

The familiar company suggestion box indicates that management wants to hear employees, but if employees later discover that their suggestions have been rejected or ignored, they come to see the suggestion box as a metaphor for corporate lip service. Unfortunately, this situation occurs far too often. Figure 2.14 is an illustration of how creativity is stifled.

KINGDOM CREATION. When the power of making the final decision resides in the hands of a few, the creation of corporate kingdoms is highly likely and easily achievable. The stereotypical television and movie view of the executive of a large company is of someone far removed from the actual business. The executive makes decisions with little regard for their impact on the employee and sometimes even on the business. The *ivory tower syndrome* reflects the idea that the large established business literally runs itself; that the goal of upper management is to attiain the "executive bathroom key."

- Stifling creativity can occur when a centrally managed company does not balance its goals with the needs of the employees
- Employee suggestions are ignored
- Managers dictate
- Employees feel as if they are considered unimportant to the company
- Employees feel as if management wants to only dictate
- Employees no longer serve as an avenue of promoting the company

FIGURE 2.14 Stifling creativity.

The funny thing is, this attitude still exists, but one should not assume that kingdom creation is running rampant in the corporate world today. There are so many external and internal mechanisms to control corporate executive behavior that even though abuses exist, they are not as frequent as one might believe. However, kingdom creation is pervasive in the startup company, usually beginning with the the founder(s) of the company.

Getting a startup company up and running is a Herculean task. Company founders run around the nation looking for investment money, hoping someone will invest in their idea. One day the founder finds money and begins the process of hiring on help. Eventually, the founder's company reaches a critical point in its life cycle and must hire experienced managers to help the company grow. Until that point, the founder has probably been overseeing the company, using younger and less-experienced employees. The need for management experience may mean hiring a company president, chairman, or an entire senior staff. At this point, the trouble begins.

Founders of startup companies have invested a great deal of their emotional and financial lives in their companies. Founders have such enormous feelings of ownership for their companies that they typically cannot accept any advice or help from the very people hired to help. This relationship would be a fascinating study in human behavior, were it possible to get past the tragic circumstances that usually arise when the founder and the experienced managers hired first clash over a decision. The founder is usually the one who starts the fight, although he might feel it was the experienced manager who started the fight because the manager chose a path the founder would not have chosen. The founder still wants total and exclusive control over all decisions in the company.

When the founder and the experienced manager clash, the employees either end up taking sides or sit bewildered over how the founder and the management team spend so much energy fighting while the company collapses around their ears. Most times, investors are forced to step in to settle the fight. Usually the investors side with the experienced managers hired by the founder. Figure 2.15 is an illustration of the relationship between the hired management team and the company founders.

In the end, the company suffers and fails. Most startups have failed because of poor management style. The conflict between the founders and the hired managers is just one factor. Unfortunately, the desire to be in control and be in charge in this case is detrimental.

POOR COMMUNICATION ARISING FROM THE CHAIN OF COMMAND. If not managed carefully and with compassion, centrally managed organizations can end up hindering interdepartmental communications. Tightly controlled organizations can engender fear. Think of it this way: if you are not encouraged to offer an opinion, then the chances are you will not want to offer one. If any employee has been reprimanded for offering a view when one has not been requested, the likelihood of ever hearing from that employee again is zero.

Even if management wants to hear from employees, a multilevel chain of command can discourage even the most sincere from pursuing any suggestions because of the sheer numbers of people involved in deciding whether or not an idea has merit. The process becomes too cumbersome. An idea may lose merit just because the environment has changed and therefore the problem has either changed to something else or gone away.

FIGURE 2.15 Founders and the hired management team.

If a senior manager in a company is responsible for a particular area, he may deliberately avoid addressing problems unless his president notices the problem. The author once heard a senior manager say "If you ignore a problem long enough it will either go away or some other person will solve it." Another senior manager once said, "It was not a problem until you told everyone there was a problem. Now we have to fix it." These managers represent the absolute worst of management. The horror is that they are in powerful positions; the other staff see these traits as ones to emulate in order to achieve success in the corporation. It is bad enough that every idea has to be funneled all the way up to the top, but to discover that no ideas make it past these senior managers for review causes the organizations simply to stop all creative thinking. If the reward for caring was criticism, then why bother saying anything at all? (Fortunately, competent and sincere managers caught up with the two walking disasters I quoted, and they were eventually removed from the company.) This example represents a combination of the worst of a chain of

FIGURE 2.16 Communication and the chain of command.

command and the worst traits of a manager. Figure 2.16 is an illustration of how the chain of command can adversely affect the flow of communication in the company.

HIGHER MANAGEMENT REPRISAL. Though we have already discussed this point, it is important enough to bear further examination. How many organizations have you worked in where you have had to think about how management would perceive a suggestion from you? How many times have you had to stop and think about how your boss would feel hearing you make a suggestion that you thought was critical to the future of the company? How often do you think about how your boss might look if he or she looked bad in front of his or her boss?

This is reality. We all stop and think about how the boss will feel if we make a suggestion. What happens if your boss thinks you are making so many suggestions that you are not doing your job? Does your boss even know what to do with your recommendation? What happens if the upper management takes your recommendation, analyzes it, and then says you are wrong? Now what do you do? Now you look like a fool; it seems that you have just wasted the time and money of the company reviewing your recommendation. Worse, the senior management's time was wasted.

What happens if senior management agrees with you, implements your idea, and you are proven wrong later? Do you think people will forget it was your idea? The answer is no. Do you think people will blame you? The answer is, of course, yes. It is the nature of people and corporations; someone always has to be blamed. Unfortunately, you will never see the punishment coming until it is too late to do anything about it. Politics, especially corporate politics, can be particularly nasty. Retribution may be slow in coming, but it will come.

The retribution does not have to be explicit. Your opinion may never be taken seriously again. You may never again be given a chance to make any major decisions in the corporation. Worse, you may be forgotten and left where you are; no one will lay a finger on you; you will simply become persona non grata. Unlike corporal punishment, which is immediate and physically hurtful, corporate punishment is quiet and rarely traceable. Are you an instructor seeking tenure in a university? Are you in line for a promotion in a Fortune 100 company? Do you want tenure or a promotion at some point in the future?

You are going to stop and think twice before you want to be a hero and open your mouth. Figure 2.17 is a rendering of how higher management can react to failure.

Though this is often forgotten, managing with compassion and understanding is key to running any organization.

DECENTRALIZED DECISION MAKING

Readers may get the impression that I am a big supporter of centralized management. They would not be totally wrong. I am, in fact, a supporter of centralized management using decentralized management principles as a balance. The danger of relying exclusively on one style of management is the possibility of stifling the creativity a company needs for growth. I will explain this idea further in this chapter.

Decentralized decision making became popular in the late 1970s. Companies sought ways to become more competitive with the then-emerging competition from countries like Japan. Decentralized decision making reduced overhead and time delays in decision making. But there are advantages and disadvantages to decentralized decision making. After they are list-

- Some senior managers will simply blame everyone else
- Some senior managers will not listen to reason
- Retribution can be slow or quick
- Employees no longer want to participate in the future of the company
- Employees no longer care about the future of the company.

FIGURE 2.17 Reprisal and higher management.

ed, I will briefly discuss each. The advantages of decentralized decision making are:

- Rapid decision making
- Facilitation of the creative process through the entire employee population
- Highly dynamic and quickly reactive to changes in the market environment

The disadvantages of decentralized decision making are that it:

- Can lead to anarchy among management and the rank and file
- Can have a negative impact on a company's overall image
- Can lead to poor communications because of the chain of command
- Can lead to bad decisions because the various business units are making decisions without sufficient information

Decentralized companies can be as vertical or flat as centralized companies. However, the key difference is that a remote or regional office, a department, or even a single employee can make major decisions without consulting headquarters. In a decentralized company one person can make a difference, be it positive or negative.

Figure 2.18 is a representation of the advantages and disadvantages of decentralized decision making.

ADVANTAGES OF DECENTRALIZED DECISION MAKING. We find that both centralized decision-making organizations and decentralized ones share the same strengths and weaknesses. Consider them to be two sides of a coin. Decentralized decision making structures have their place in large companies, small companies, and even startups. The key is to know how to apply the strengths of these models or structures. It is necessary to balance management structures in a way that optimizes the strengths of the employees.

The advantages of decentralized decision-making are:

- Rapid decision-making
- Facilitates the creative process
- Highly dynamic and reactive to changes in the market environment

Decentralized decision-making based companies tend to be flat organizations

The disadvantages of decentralized decision-making are:

- Can lead to anarchy among management and the rank and file
- Can impact a company's overall image
- Can lead to poor communications
- Can lead to bad decisions because the various business units are making decision without sufficient information.

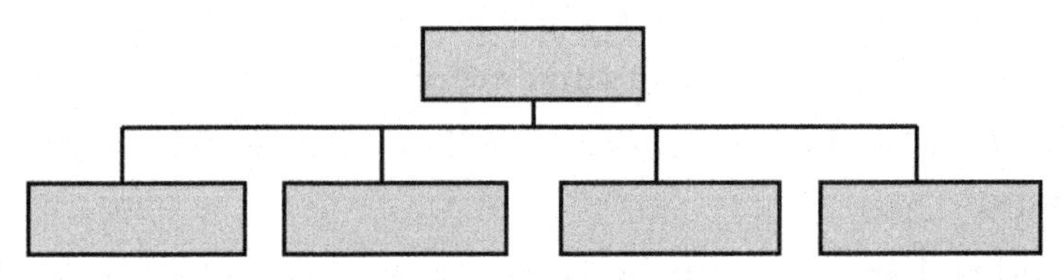

FIGURE 2.18 The advantages and disadvantages of decentralized decision making.

RAPID DECISION MAKING. In the previous sections on centralized decision-making structures, I noted that decisions are normally reached far more quickly in a centralized-authority company. I also noted that this situation is achieved because of the small number of people needed for support. The same can be said of a decentralized management structure. The key point is that the remote or regional office must be in sync with the management plans and the philosophy of headquarters.

Having headquarters make every decision can be very time consuming. However, if the remote or regional offices are allowed to operate as autonomous entities without control, then the ensuing anarchy created by regions' overstepping each other will be beyond imagination. Often headquarters tries to handle those issues that affect the entire company. This would include approved vendor lists and marketing programs. The smaller issues are left to the local or regional offices to manage.

Only regional offices can react quickly enough to specific regional market needs. Figure 2.19 is an illustration of the

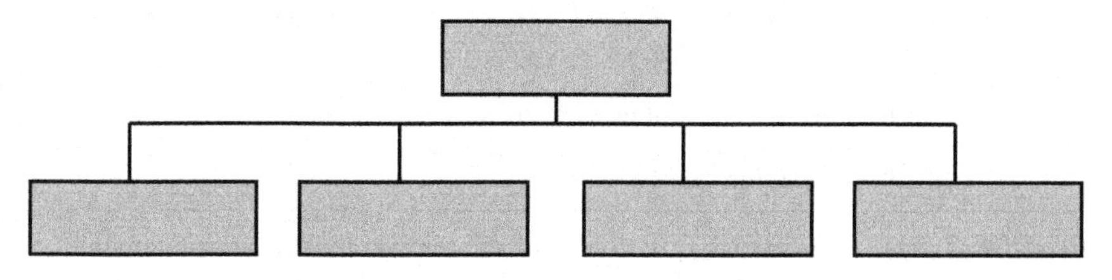

- Decisions can be made quickly because there is less headquarter involvement
- Fewer management people are involved in the decision
- Employees are encouraged to make suggestions

FIGURE 2.19 Rapid decision making in a decentralized management environment.

impact of the decentralized decision structure on the rapidity of decisionmaking.

FACILITATING THE CREATIVE PROCESS THROUGH THE ENTIRE EMPLOYEE POPULATION. When employees feel they can voice their views freely and that those views will be seriously considered without the possibility of political retribution, then a company will discover it has turned on the creative energies of all its employees. The challenge for management will be to manage these energies in such a way that creativity is properly harnessed, which means a structured way to review all the suggestions in an unbiased manner.

Creativity in a telecom company that provides service is usually focused on finding new ways to keep customers from leaving and buying service from another carrier. Creativity in technology is involved to the extent that it is focused on creating a service that supports the carrier's sales and marketing efforts. Creativity in a telecommunications hardware or software company is focused on creating a product that the customer, usually the carrier, will buy. The customer is the focus of all these activities. Figure 2.20 illustrates the focus of a telecommunications company.

To focus efforts properly, there must be one department responsible for the management of all product concept development. In both service provider and vendor shops, product ideas are managed in the following manner. The sales department obtains input from the customer. Marketing takes this

- Customer or users pay for service
- Customer or users are the focus of every telecom service provider
- Profit is not the only company driver

FIGURE 2.20 The customer: the focus of the telecommunications company.

input and market research data on a customer segment and sends it to a product management group and a product planning group. The product planning group takes this input and conceptualizes the product line to be developed. Product management takes this concept and brings it to reality. Note that the product planning function is one that oversees all activities in a company in order to ensure alignment with the company's business objectives for the year and long-term activities. In some companies the product planning function resides within marketing, while in very large companies, the product planning function resides in a separate department that reports to the president and CEO of the company. Product planning is a strategic organization that is key to setting the direction for a product line and conceptualizing new products.

The product management group collaborates with the product development (technology) department and creates the product. The product management group, which usually resides within the marketing department, is responsible for translating customer input into achievable technology requirements. While the product is being made, there is constant communication between all departments in order to ensure alignment with changing market conditions and changing customer needs. Remember, as time passes, the market and the customer will change their views. The product creation process will be considered more fully later in the book, but the point is that there is a series of steps and a process of organization involved in the product creation process. Further, there is one organization responsible for overseeing the creation of the product. Note that there is communication between people and yet there is structure. This is the way creativity is encouraged yet managed. Figure 2.21 is an illustration of the product creation process.

Creating products and services is a process that requires an understanding of the current and future needs of the customer.

- Customer input is essential
- Create product that people will pay money for
- Listen to the customer
- The customer is a business-critical element of any business

FIGURE 2.21 Creating products and services.

To reiterate, the process of product and service creation is divided into tasks and responsibilities. All departments participate and work together. People are required and encouraged to make decisions and speak up.

HIGHLY DYNAMIC AND REACTIVE TO CHANGES IN THE MARKET ENVIRONMENT. When a company encourages creativity properly, this should lead to an operating environment that encourages nimbleness of thought. In other words, an environment of teamwork. Theory and practice are two different things. When you introduce the "personality factors" (corporate politics and ego), achieving teamwork becomes an ongoing struggle.

Teamwork is the only way a company can attain the level of capability needed to respond to rapid market and customer changes. This teamwork must be focused on meeting customers' needs and ultimately generating profit. In an ideal world, teamwork without the effect of employees' personal agendas would make for a powerful company. Imagine a company where the collective talent of all employees is harnessed for one goal.

A company that is decentralized in its decision-making process must, by default, trust its employees to make decisions and carry out tasks for the greater good. Once this trust is properly organized and harnessed, we have an organizational structure that can proactively respond to nearly any change in the marketplace.

DISADVANTAGES OF DECENTRALIZED DECISION MAKING. No system is perfect. Decentralized decision making is a proactive view of dealing with employees and a positive method of delegating decision authority. However, as in any system, there is a down side to the decentralized decision-making process. As I noted in the previous section, the disadvantages of decentralized decision control processes are that they:

- Can lead to anarchy among management and the rank and file
- Can have a negative impact on a company's overall image
- Can lead to bad decisions because the various business units are making decisions without sufficient information
- Can lead to poor communications because of a breakdown in the chain of command

Figure 2.22 is an illustration of the above points.

ANARCHY IN THE COMPANY. It is critical that this point be understood: when you have a management structure that is decentralized in nature, you are inviting a large number of people to participate in the policy and decision-making processes of the company. You run the risk of the "too many cooks in the kitchen" syndrome. In other words, too many people offering views where none are needed increases the risk of decisions made because of personal agendas. This increases the likelihood of dissension among management and confusion among the employees.

In my own experience, egos get in the way once low-level managers are given the ability and authority to make even low-level company decisions. The low-level managers given this authority suddenly decide that the senior management is

The disadvantages of decentralized decision-making are:

- Can lead to anarchy among management and the rank and file
- Can impact a company's overall image
- Can lead to poor communications
- Can lead to bad decisions because the various business units are making decision without sufficient information

Everyone is in charge and no one is in charge

FIGURE 2.22 Disadvantages of decentralized decision making.

wrong about the direction of the company. Rather than discuss the issue with executive management, the managers decide to undertake independent action without permission. When you have managed people for a long enough time, you see both good and bad managers. When you invite multiple people to participate in a certain kind of decision, you increase the risk of hurt feelings and dissension when the decision does not go as managers wish.

This does not mean it is not possible to build a consensus. Building consensus is a management technique used to increase cooperation among management staff without giving all the managers the ability to say "yes" or "no." Consensus building is a technique often employed in companies structured on a decentralized decision-making model. Consensus building is a way to ensure that the entire management staff is aligned with the goals and plans of the company.

The problem for the inexperienced manager building a consensus is that often the activity is confused with making the final decision. Building a consensus does not entail getting total agreement on every issue. Rather, it is about reaching a general agreement. Consensus does not mean agreement on every detail. Too often, I have seen corporate presidents go overboard and seek complete agreement as opposed to general agreement; this eventually leads to a confused management team. The confusion occurs because the management team suddenly believes its views are essential on every issue. In fact, what needs to be done is to ensure that managers understand their opinion is being sought, but that the final decision is reserved to one manager (the senior manager or president). The inability to communicate this last message via action and words creates an environment that leads junior managers to believe their total agreement is essential. This mistaken belief will lead these junior managers to feel they are "making the decision" as opposed to participating in educating the senior manager and socializing the issue among their peers. The senior manager can create an atmosphere of ill will, once the decision is made and is different from what the junior manager believed it should have been.

We need to remember that management is not all science; it is mostly art. We will investigate this subject and the above scenario later in the book. Figure 2.23 is a representation of the discord created by improperly executing a decentralized decision-control process.

Decisions not properly communicated through the work force can lead to confusion and discord amongst the employees.

Consensus management not executed correctly leads to stagnation.

Decentralized decision making processes can lead to many bosses.

FIGURE 2.23 Discord in the work place.

ADVERSE IMPACT ON A COMPANY'S OVERALL IMAGE. Activities within a company can affect the public image of that company. One way of looking at this issue is that most people cannot hide their own feelings when things are not going well at home. A company cannot hide the feelings and attitudes of its employees when things are not going well for them at work. This does not involve the financial performance of a company, but rather, this issue is about how the employees' feelings about the company are communicated to the public.

How often have you complained about your boss or your company? How many times have you complained about an employee? How often have you made your complaints known to people outside your company? How often have you done your job and no more than your job because you just could not stand the sight of your boss or you hated your own company?

Bad feelings as well as good feelings about a company are always communicated via employees through words, action, inaction, and appearances. When a decentralized organization does not manage the decision-making process in the company well, an atmosphere of ill will is created among employees. Figure 2.24 illustrates how internal discord is communicated to the public.

Decisions not properly communicated through the work force can lead to confusion and discord among the employees

- Employee frustration and anger eventually is seen by the customer
- Sometimes the customer is victimized by the disgruntled employee

Customer care is a business-critical area of a telecom service provider

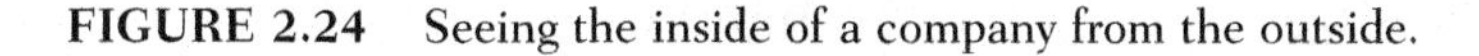

FIGURE 2.24 Seeing the inside of a company from the outside.

BAD DECISIONS WHEN THERE IS INSUFFICIENT INFORMATION. A decentralized decision-control process that has not been properly established can result in situations in which information is not being disseminated to all appropriate parties. When this occurs, bad decisions are made because the managers had insufficient information.

Many managers will tell you that they have made decisions when sufficient information was not available. Often managers are forced to make decisions without all the facts. That is fine—it's part of reality. But we should not invite the problem to our doorsteps. Managers must strive to gather as much data as possible in order to make a decision. They must have as much data as possible made available to them in order to do their jobs properly. Decentralization not only decentralizes the decision-making process, it also disrupts the flow of information. In other words, the dissemination of information suffers from misdirection and dilution. Disseminating information to the right people becomes a challenge.

Questions that come to mind are: How do you decide what information needs to get to the correct manager? How do you get the information to that manager in a timely fashion?

It is always possible to disseminate all information to all managers, but that establishes an environment in which every manager is invited to participate in the decision. Furthermore, irrelevant information becomes a distraction to those managers who do not need it. In this case, it means more email or letters for the manager to sort through.

COMMUNICATIONS BREAKDOWN DUE TO A BREAKDOWN IN THE CHAIN OF COMMAND. A decentralized decision-control process can lead to enormous confusion among the various levels of management because no one knows who is in charge. Someone must be held accountable for bad as well as good decisions. The decentralized decision-making process has often been used as a smoke screen by incompetent managers to cover their own inadequacies. I call this situation the "not my job" syndrome.

A decentralized process does not mean there is no structure for decision making. Rather, this structure involves more than just a single individual or handful of managers. The decentralized decision-control process is a structured and defined organization that involves many people. When the process is not understood by the employees, then bad decisions are made and good decisions that are made may not get communicated to the right managers. When the latter situation occurs, there is a complete breakdown in the management chain of command. Figure 2.25 is an illustration of this type of breakdown.

WHAT IS THE MODEL TO FOLLOW FOR TELECOMMUNICATIONS COMPANIES?

Where is this all leading? Both centralization and decentralization have their pros and cons. What organizational structure is best? The model to follow depends on the type of business being addressed. However, the kind of industry you are in is only part of the answer. The type of business model the company is following is also a factor.

In the telecommunications industry, the business model is a major factor in how the company should be structured. Depending on the type of telecommunications company, the market segment the company is servicing, and the geographic

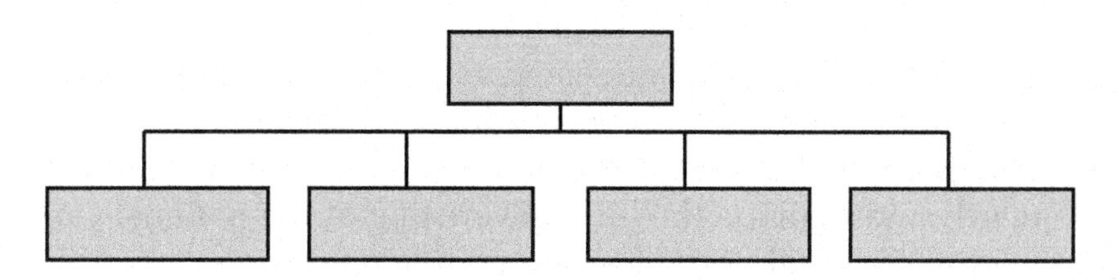

When the decision making process is not understood, the entire decision making process breaks down.

- Decentralized processes have many points of failure.
- When a decentralized management organization breaks down, the breakdown spreads like a brush fire out of control. Those seeking more autonomy take advantage of the confusion.

FIGURE 2.25 Chain-of-command breakdown.

area being covered, the company may be centralized or decentralized in its management structure.

In most industries, when a company is centralized in its management structure the company's main operations are geographically focused in one area. In most industries, when a company is decentralized in its management structure, the company's operations are geographically dispersed throughout the nation or a region. However, the telecommunications industry is not like most other industries.

The telecom industry is comprised of carriers and vendors of hardware and software. It is characterized by nearly instantaneous service, rendered between carriers and customers, between vendors and carriers, and between vendors and vendors. Service providers (carriers) like to deal with vendors who have local representatives in each market.

Carriers have representatives in every area in which they operate. National carriers can have dozens of offices in a single market to address customer needs; these offices are called the *customer service centers*. National carriers have field forces "on the ground" in every local market because the carriers have switching equipment, radio equipment, or other types of telecommunications equipment in every local market.

Whether a national telecom carrier follows a centralized model or not, it will have local marketing and sales forces in every local market. A person sitting in the state of New York can create a national image, but that person cannot create a local message that communicates to the local market in the proper local language and nuance. Unless you travel across the country frequently and examine the local advertising of the national carrier, you might think the television commercial you are looking at in your area is the same one seen on the other side of the country. It is not; even a national carrier will localize a television commercial with reference to the area and even use local slang (language and terms used by those in a local area).

In regard to the sales and marketing of telecom services to users like you and me, telecommunications services provisioning is like national politics: at the end of the day it is all local—most national politicians still must gear their campaign stops to those issues that motivate the voters just as telecom companies must target local consumers. As prominent as the majority leaders of the United States Senate, the United States House of Representatives, or the Canadian parliament may be, they still must be voted in by their local constituents. Even the President of the United States has to gear his remarks to those issues that are not only national, but of concern to the populace he is speaking to. Telecommunications service provisioning is the same: no matter how big the carrier; the advertising, the service offering, and price must appeal to the local population.

Can a centralized company achieve this level of communication locally without some amount of local knowledge? The answer is no. Furthermore, the only way to achieve this local knowledge and incorporate it into the company is through some amount of decentralization. Figure 2.26 summarizes the issue of centralized versus decentralized telecom organizations.

As you may have guessed, I support both centralization and decentralization. Having such a position is not a sign of indecision. The optimal organizational structure for the telecommunications carrier or vendor is one that combines the best of both.

SERVICE PROVIDER DECENTRALIZATION GONE WRONG

In the late 1980s and early 1990s, cellular carriers were managed via a decentralized decision-control process. Each region

You cannot sell to someone unless you know them.

- Selling requires knowing the market.
- Knowing the market means knowing the people in the market.

Telecommunications services is like national politics: at the end of the day it is all local.

FIGURE 2.26 Local knowledge importance.

was managed by a general manager and had its own engineering, operations, sales, and marketing organization. The regions considered headquarters a necessary evil with which the regulatory bodies and public utility commissions interacted.

Cellular carriers used to let local general managers run their regions with their own set of rules, their own switching equipment, their own human resources departments, their own hiring policies; they could manage network interconnection with the Incumbent Local Exchange Carrier (ILEC) their own way. In fact, between 1984 and the early 1990s each cellular carrier looked like multiple carriers residing within a carrier. This was the epitome of decentralization. Things became so bad that as cellular carrier customers roamed from region to region, they encountered problems with their service. What the customers discovered was that the regions were not selling the same kinds of services. In other words, you could be in New York and then travel to Rhode Island only to discover that the same carrier did not have its network provisioned to provide voice mail or call waiting. If the two different states did provide the same services, you could only access it using different dialed digits. Sometimes the regions did not have their databases synchronized: the customer's handset could not register in another state because it did not exist in that state's database. Operational procedures were different. Technology was different. It was so bad for customers that every trade journal at the time published articles about these various customer issues.

It was not until the early 1990s that the large cellular carriers got their acts together and centralized their overall operations and policies. Needless to say, many employees complained about this sudden big-company policy making and behavior. This behavior was necessary because many cellular carriers had reached a size that optimizing their processes was necessary for the companies' survival. In fact, activities considered normal for the wireline carriers were foreign to the cellular carriers. The activities optimized through centralization were:

- Procurement
- Marketing
- Sales
- Finance
- Technology
- Business development
- Public relations
- Operations

In a centralized model, the regions are still responsible for all of the above but with overall direction from headquarters. Therefore, the regions are responsible for providing local marketing spin and control of their region. However, the key point to remember is that headquarters sets and enforces the guidelines within which the regions operate. Figure 2.27 is an illustration of this centralized/decentralized operation.

One thing that has not changed in the carrier community is fiscal accountability and profit and loss: the regions are still responsible for making money.

Telecom manufacturers of software and hardware are no different from the carriers when it comes to dealing with organizational issues. The hardware and software companies have the same functional organizations as do carriers. The activities of these organizations differ in specific tasks, but these companies are still making, marketing, and selling products. However, the impact of decision decentralization on manufacturers has not been publicly advertised, as it has for carriers. Nevertheless, hardware and software companies have encountered the same organizational problems as those faced by carriers. Those in the telecom community have intimate knowledge of the problems created by vendors' organizational issues. The problems include mis-sized product, wrong price, miscommunicated customer needs, insufficient manufacturing capacity, software bugs, products with the wrong capabilities, and even products delivered to the wrong location.

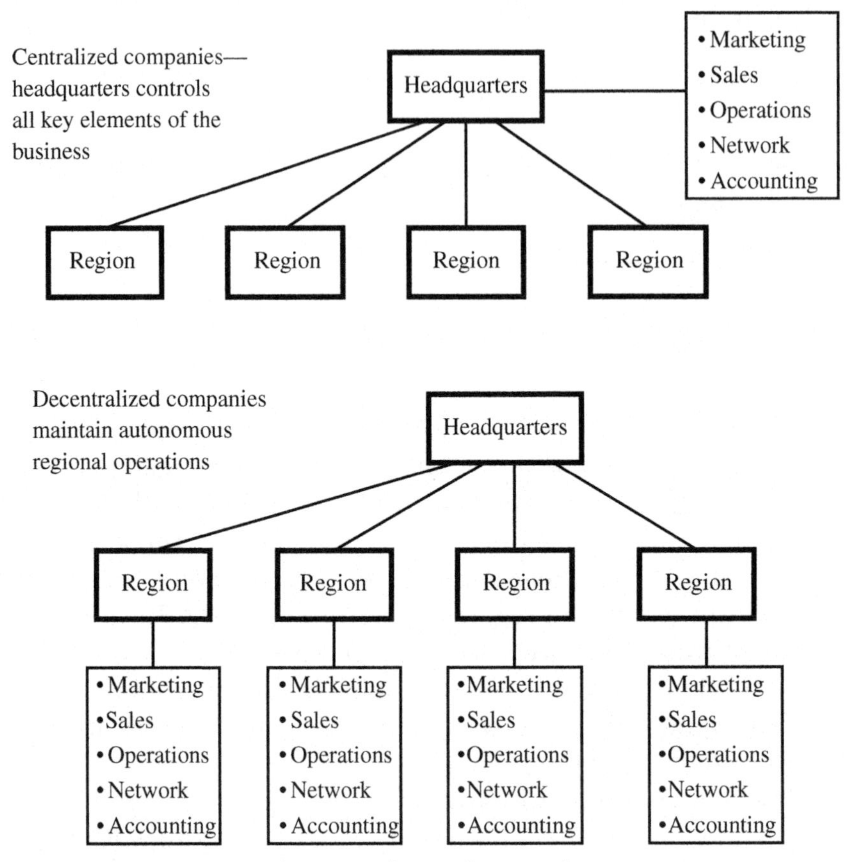

FIGURE 2.27 Decentralization and centralization of a carrier.

OPTIMIZING THE ORGANIZATIONAL STRUCTURE

One of the most widely cited models of decentralization has been the General Motors Corporation. The late Alfred P. Sloan, former chairman of General Motors, was one of the creators of the model currently used by cellular carriers, wireline carriers, and even manufacturers. This model focuses on centralization but it has a decentralized component as well. The Sloan model supports two premises:

- The chief executive of a department or organizational leader shall be responsible for his own operation. This responsibility is total. The chief executive of the organization shall be empowered to exercise full authority to operate and make decisions to enable the operation to achieve its goals. This is decentralization.

- Certain central organizational functions are necessary to the development of the corporation and the appropriate coordination of the corporation's activities. This is the function of central staff, also known as headquarters.

Sloan created a model known as Centralized Control of Decentralized Operations. Most large organizations have patterned themselves after the Sloan model. The Sloan model is a balance of the best of both worlds. Large companies today are global, with multiple operations and tens of thousands, even hundreds of thousands, of employees around the world. Figure 2.28 is a representation of the Sloan model.

The Sloan model enables corporations to overcome potential breakdowns in communication, control, and managerial effectiveness. If applied properly, the Sloan model will address

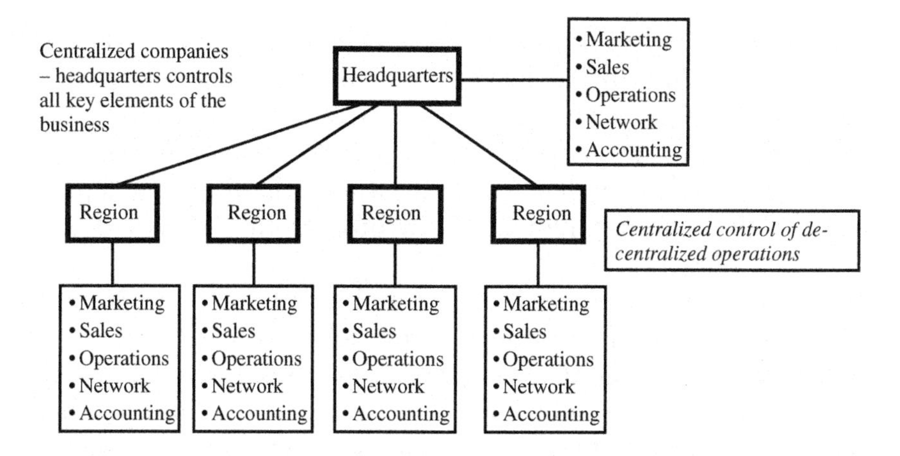

FIGURE 2.28 Sloan model.

not only the big management issues but also day-to-day operational issues. The routine day-to-day benefits of applying the Sloan model include:

- Rapid decision making
- Clearly defined roles
- Absence of conflict
- Organizational politics kept under control
- A management structure that encourages suggestions and even criticism
- Reduction of the perceived privilege gap between the senior management and the overall population
- Weeding out weak managers; weak managers are also kept from riding on the coat tails of the successful managers
- Facilitating the creation of a policy-driven corporate management structure, rather than a management structure that is based on the whims and capriciousness of senior management

These attributes represent the ideally managed company. (I am not discounting decentralized management structures. Decentralization is very important when one is managing a company with a diversity of assets that are geographically dispersed or a company that has its multiple organizations/departments. One cannot manage everything from the top; it just does not work.)

Telecommunications carriers and some telecommunications manufacturers (hardware and software) are structured around the Sloan model. However, this model is not the only management model, although it is a popular one and has had the longest period of success. I strongly support this model.

Theoretically, when you create a decision-control process that is decentralized, you are creating a management process focused on one of two things: managing physically dispersed operations or delegating of decision-making authority. When you create a centralized management structure, you are focused on ensuring uniformity of decision making and rapid decision making. However, as with anything in life, to do too much of anything is not good. Life is about balance. Management is not just about managing processes, it is about motivating and working with people. Leadership is not about "leading the charge up a hill" to defeat the enemy, but about leading a charge accompanied by forces who want to charge the hill with you because they don't want you to go into battle alone. I draw a distinction between the two because that is reality. Great companies have great employees, great managers, and great leaders.

The creation of a management structure requires striking a balance between various components of the decision-control processes. These processes alone cannot ensure a company's success: they are tools. Tools coupled with the skills of experienced managers and leaders are required to ensure the success of a company. To strike a balance between all the various business models is an ongoing challenge for any company. Figure 2.29 is an illustration of the balance that is needed.

SUMMARY

Telecommunications companies in the United States have seen tremendous change since 1984. Globally, the telecom industry has undergone the same kind of market changes for just as long. These changes have caused many telecommunications companies to disappear from the business landscape. The telecommunications market has changed so much from the early 1980s that many telecom professionals from that period cannot even recognize the industry today. Those who have survived and stayed in the business have had to adapt by either learning new skills or adapting their existing skills to the new environment.

Telecom managers and other professionals have applied various management and leadership theories in varying degrees and combinations with varying levels of success. To criticize

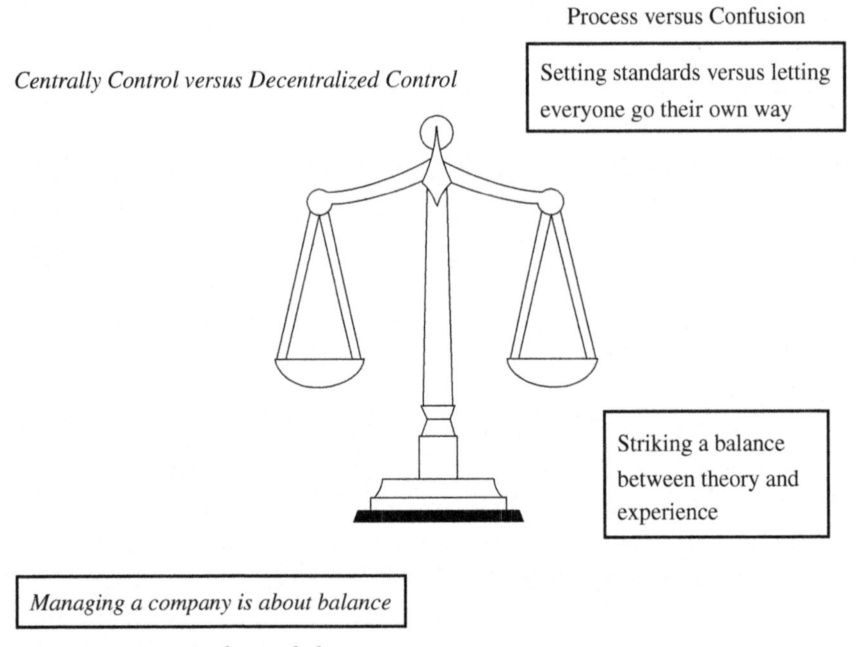

FIGURE 2.29 Striking a balance.

the management style of one professional is difficult unless you are willing to have the spotlight shone on yourself. However, criticism is very easy when a telecom professional manages without compassion for the workforce. Despite the varying theories of management, one thing holds constant for all the theories: lead and manage with compassion.

To a great extent, the success of management theories has depended on the personalities of the management team and the respect they have had for each other and the employees who support them. To manage without compassion is wrong.

Many years ago a mentor told me that one day I "would have to order 100,000 people to do the impossible. Those people would only do the impossible for someone they respect, so show them the respect they deserve." That was the best lesson in leadership I have ever had.

This book is not about leadership but throughout it, keep in mind that people are involved in all aspects of management. You cannot treat a company like a set of numbers, without taking the people into consideration. Telecom companies, especially the carriers, serve the public and sell services to the public. How they are managed is reflected in the customer base. The next chapter takes a look at how telecom service providers (carriers) are organized and managed.

CHAPTER THREE

THE TELECOMMUNICATIONS SERVICE PROVIDER

As I noted in Chapter 1, a provider of service is also generically called a carrier or service provider. The user or purchaser of the service can be an individual or a company. The different kinds of carriers include:

- Wireline telephone companies
- Cellular carriers
- PCS (Personal Communications Service) carriers
- Paging companies
- Cable television companies
- Satellite carriers
- Internet service providers

Despite the different technologies and market segments being served, they have one thing in common; they all provide service. Furthermore, telecommunications service providers are all regulated by government to some degree.

Telecom service providers are like any service companies. They are comprised of an employee population with skill sets totally focused on ensuring reliable service to the customer. Figure 3.1 illustrates this customer focus.

- The customer is the focus of the telecom company.
- The customer is the source of the revenue.
- The customer's needs must be anticipated.
- The customer's needs must be fulfilled.

FIGURE 3.1 The customer focus.

BASIC ORGANIZATIONAL FUNCTIONS

The types of skill sets or functions needed to provide telecommunications service fall into a number of different organizational areas:

- Operational/maintenance
- Technical
- Administrative
- Management
- Business—sales and marketing
- Finance
- Human resources

The categories are very broad and are applicable to any company. Organizational management is an area of knowledge applicable across any industry. All telecommunications carriers maintain the same functional areas with differences directly linked to the creation of the service. Figure 3.2 is an illustration of the basic organizational functions of a telecom carrier.

I draw a distinction between organizational functions and organizational structures. The functions I describe do not really shed any light on how the company is structured from a reporting-structure perspective.

Those managing or investing in a telecommunications carrier can discern a great deal when they examine the organization chart that lays out the entire decision-making process. Understanding the types of functions needed to run a carrier facilitates examination of a carrier with a view to investment.

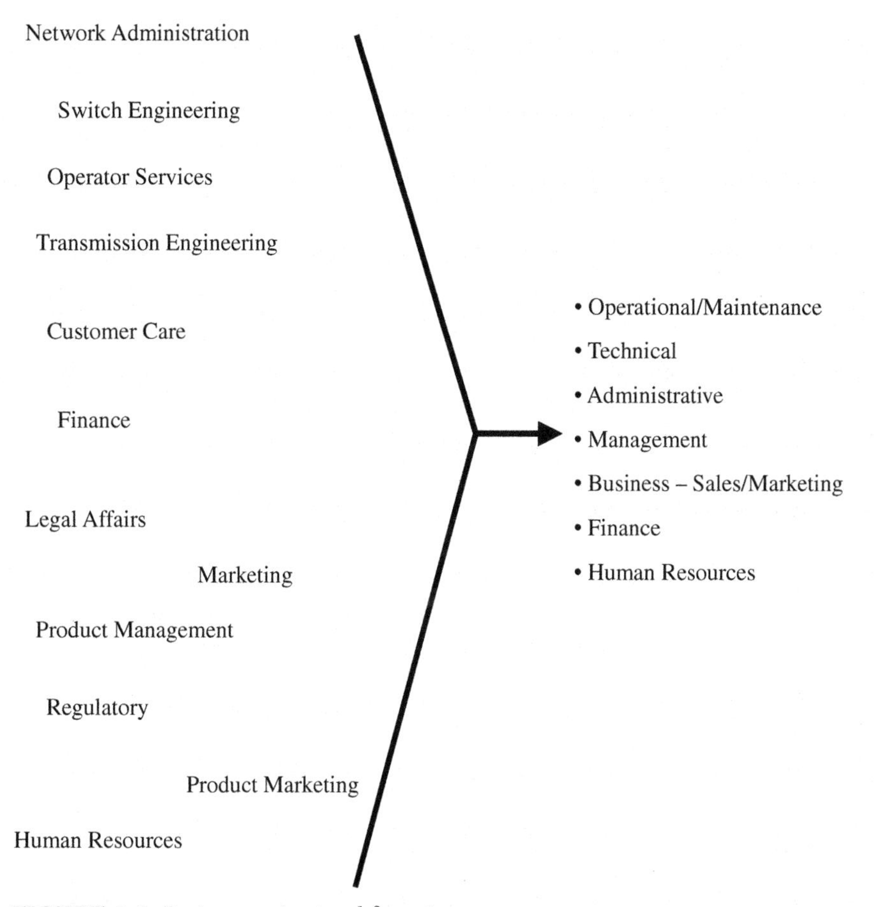

FIGURE 3.2 Basic organizational functions.

No matter what type of carrier we are addressing, the functions are the same. If organizations did not behave or function in very similar ways, then university degrees in business would be useless in the real world. Even lessons learned the hard way, by mistakes made in a real operating environment, would not be transferable from business segment to business segment. Many universities generate business for themselves simply by analyzing how companies behave and publishing papers about those companies.

However, theory is no substitute for experience. Every industry has its own culture, language, and tools. In some instances, experience is more important than book knowledge.

Experience is a governor or rudder for actions and decisions; it can enable us to make decisions more quickly than if we did not have it. If organizations did not function or were not structured in similar ways, then unemployment would be far worse than it is today. Skills learned in one company can usually be transferred to other companies.

The following three subsections will consider the telecom industry's three major segments: wireline, wireless, and Internet. We will take a quick look at the basic functions of each segments. Then we will address organizational structures, decision-making control processes, and planning processes.

WIRELINE TELECOMMUNICATIONS CARRIER ORGANIZATIONAL FUNCTIONS

Specific organizations and subgroups in a wireline telecommunications carrier are comprised of a set of basic functions:

- Corporate staff
  - Strategic planning
  - Program management—large-scale projects usually corporate-wide
- Corporate medical
- Human resources
  - Benefits administration
  - Employee issues (such as employee harassment)
  - Affirmative action
  - Hiring policy administration
  - New hire administration
  - Employee termination
  - Code of conduct administration
- Corporate security—All regional security reports directly to this organization
- Marketing
  - Product marketing

   - Product management
- Sales
   - Residence
   - Business
   - Large accounts
   - Small and medium-sized accounts
- Transmission (outside plant) engineering
   - Design
   - Equipment engineering
   - Procurement and installation planning—Sometimes transmission systems are leased from other carriers
- Switch engineering
   - Applications/services development
   - Design
   - Traffic study
   - Equipment engineering
   - Procurement and installation planning
- Finance
   - Taxes
   - Internal accounting
   - Audit
   - Revenue assurance (sometimes a separate organization with its own vice president)
- Legal
- Regulatory
   - Lobbying
   - Filing—Rates and policy positions
   - Compliance oversight
- Repair
   - Customer—onsite with the customer
   - Field—overhead wire, underground wire, and all other repair in non-customer and non-telephone company owned

property. In many wireline companies, this group will also manage the installation of new telephone poles.

- Customer care and provisioning
  - Business account billing
  - Carrier access billing
  - Residential billing
- Service order processing
- Customer complaints
- Network operations
  - Network management
  - Switch repair
- Building and facilities construction and management
  - Real estate management
- Public relations
  - Advertising
  - Press/media management

Figure 3.3 is an illustration of a wireline carrier's organizational functions. It is important to note that every one of the aforementioned functions is focused on ensuring that the customers' needs and the corporation's operational objectives are met.

WIRELESS CARRIER ORGANIZATIONAL FUNCTIONS

Most of the organizations in a wireless carrier are the same, except for differences in the sales (individual sales replaced residence sales), engineering repair, real estate, and customer management areas. In the engineering department, instead of transmission, there is radio design:

- Radio design
  - Equipment engineering
  - Procurement and installation planning
  - Frequency planning

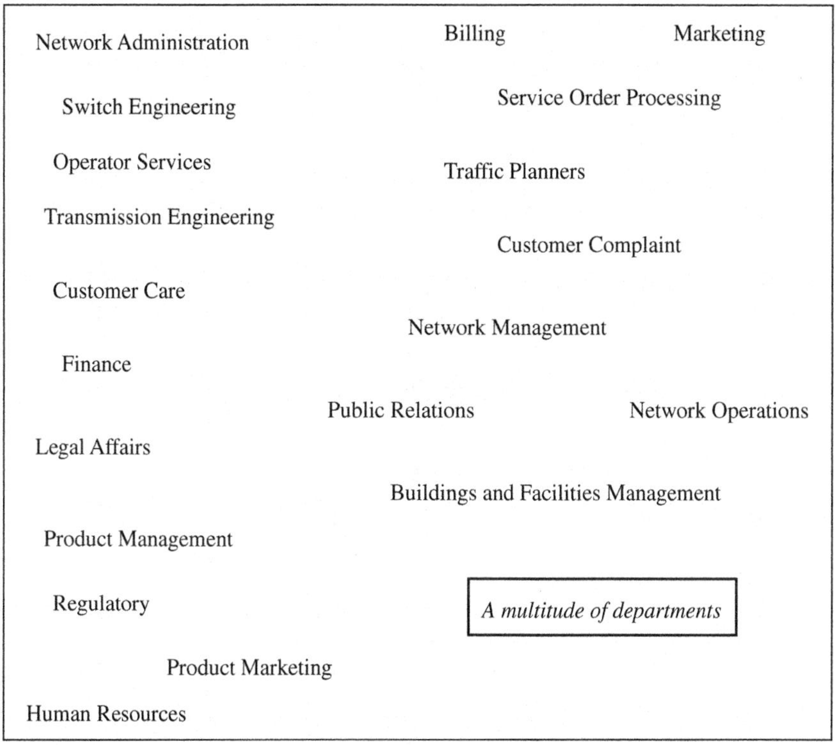

Network Administration	Billing	Marketing
Switch Engineering	Service Order Processing	
Operator Services	Traffic Planners	
Transmission Engineering	Customer Complaint	
Customer Care	Network Management	
Finance		
	Public Relations	Network Operations
Legal Affairs		
	Buildings and Facilities Management	
Product Management		
Regulatory	*A multitude of departments*	
Product Marketing		
Human Resources		

FIGURE 3.3 Wireline carrier organizational functions.

- Traffic study
- Antenna and tower siting

In customer service, user billing replaces residential billing. In building and facilities construction management, real estate management includes:

- Building and facilities construction and management
  - Real estate management—Includes tower leases as well as buildings

In a wireless carrier, the transmission department does not exist. Rather all wired or fiber optic transmission systems are leased or purchased from either the local telephone company or other carriers. The customer segment is not divided into residential and business, but into individual and corporate accounts. Furthermore, there is no repair department. Figure 3.4 is a representation of the wireless carrier's organizational functions.

INTERNET SERVICE PROVIDER ORGANIZATIONAL FUNCTIONS

In an Internet company the same kind of business functions exist. The major differences are in the technology departments. Router/Gateway systems repair is a function of network operations. Related departments include:

- Router/Gateway systems engineering
  - Design

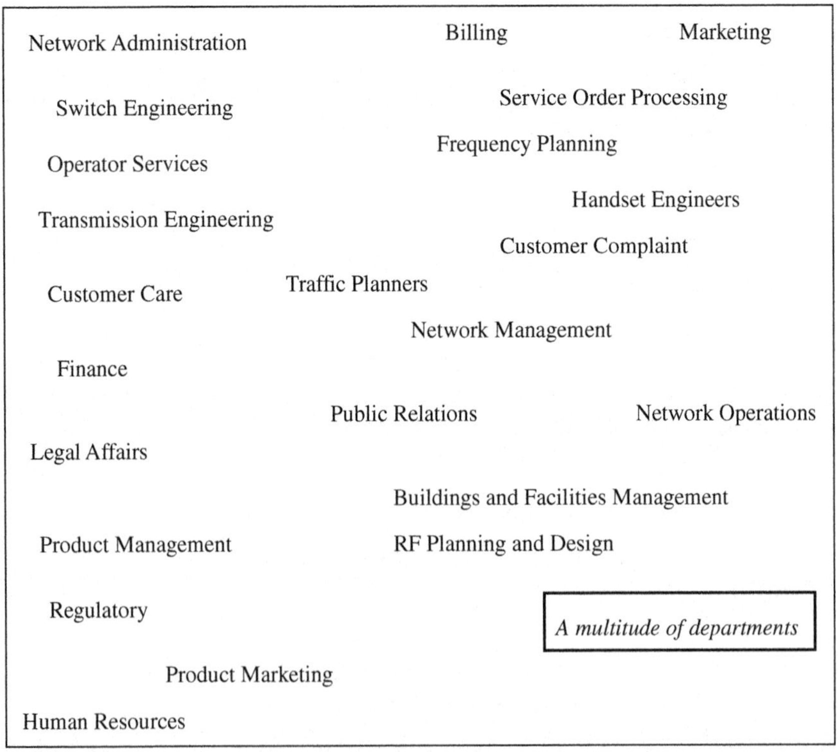

Network Administration
Billing
Marketing
Switch Engineering
Service Order Processing
Frequency Planning
Operator Services
Handset Engineers
Transmission Engineering
Customer Complaint
Customer Care
Traffic Planners
Network Management
Finance
Public Relations
Network Operations
Legal Affairs
Buildings and Facilities Management
Product Management
RF Planning and Design
Regulatory
A multitude of departments
Product Marketing
Human Resources

FIGURE 3.4 Wireless carrier organizational functions.

  - Traffic study
  - Equipment engineering
  - Procurement and installation planning
- Application/services development

Internet carriers require a great many software-related skill sets. Further, unlike the wireline and wireless carrier communities, Internet service providers do not rely on their infrastructure vendors to develop new services. This fact will be considered later in the chapter.

Figure 3.5 is an illustration of an Internet carrier's organizational functions.

Network Administration
Billing
Marketing
Router Engineering
Service Order Processing
Network Operations
Applications Developers
Transmission Engineering
Customer Complaint
Customer Care
Network Management
Finance
Public Relations
Legal Affairs
Buildings and Facilities Management
Product Management
Regulatory
A multitude of departments
Product Marketing
Human Resources

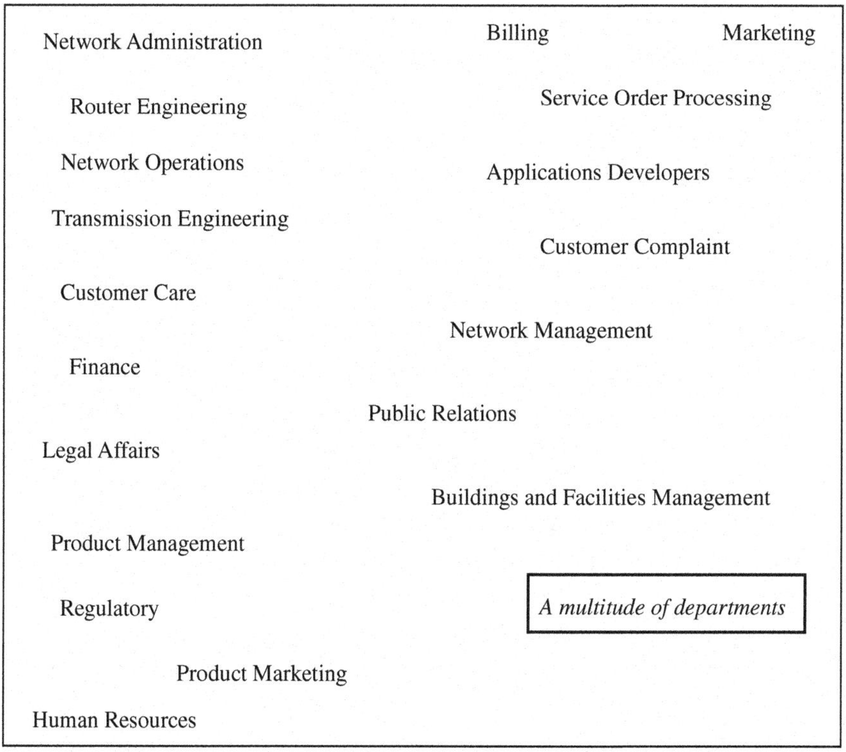

FIGURE 3.5 Internet carrier organizational functions.

CARRIER ORGANIZATIONAL STRUCTURES

Without having to read lists for every other type of carrier, you should now understand that every telecommunications carrier behaves and is organized similarly. This makes sense, because these entities are money-making, profit-making companies that provide a service. These are not charity organizations. They are focused on creating monetary value. No matter how one wishes to "slice the pie," it is still a pie. Carriers are businesses.

Functional analyses can be applied to all companies. The major differences usually lie in the product creation portion of the company, which is either the technology area or the product or service management area. We focus next on how a telecommunications service provider can be organized. We will address a number of organizational configurations. This chapter will also consider specific functions that tend to be overlooked because they are not considered "glamorous."

There are two basic organizational configurations carriers use: vertical and flat.

VERTICAL AND FLAT ORGANIZATIONAL STRUCTURES

The terms vertical organization and flat organization describe a company's spans of control and management levels. The classic principle of *span of control* was concerned with the number of subordinates a manager could effectively manage. The concept of vertical and flat organizational theory is concerned with the overall structural relationships within a company.

Figure 3.6 describes the vertical and flat organizations. The key difference is that the vertical organization has narrow spans of control or areas of responsibility, but large staffs. Notice that there are multiple layers of management in a very specific function. In contrast, the flat organization has wide spans of control. In both cases, the implication is that there are many people.

The classic bureaucratic company is vertically organized. In a vertical company there are multiple layers of managers, each with a specific area of responsibility. In a flat company there are only a few layers of management. The total number of

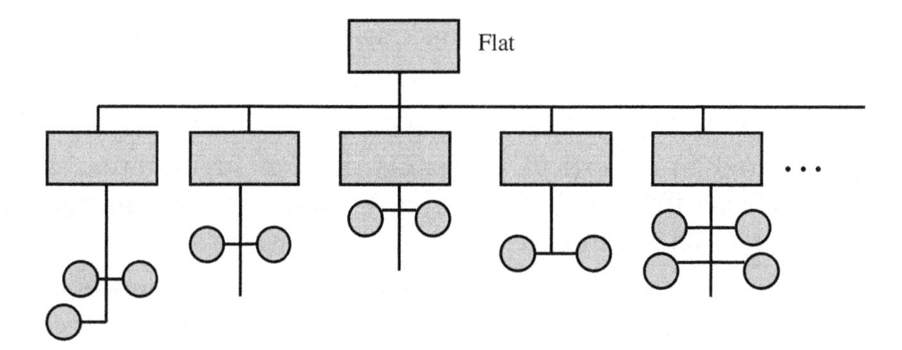

Flat organizations have fewer layers of
management than Vertical organizations

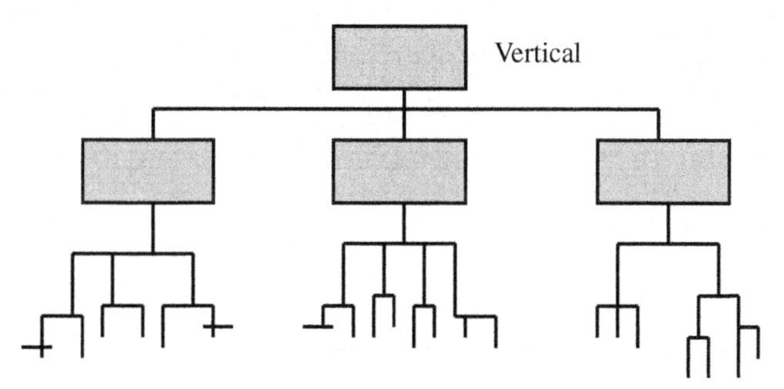

FIGURE 3.6 Vertical and flat organizations.

employees may not differ, but in a typical flat company there are usually fewer managers and therefore fewer employees overall.

Vertical companies tend to be very complex because of the sheer number of departments, organizations, and groups one must know and understand in order to properly perform one's own job. A flat company is less complex, with fewer organizations, departments, and groups to deal with. The flat company tends to give individuals more responsibility and encourage personal initiative.

Telecommunications carriers can be both flat and vertical. The Incumbent Local Exchange Carriers (ILECs) are vertical, albeit not as much as they were 25 years ago. Most Competitive Local Exchange Carriers (CLECs) are flat. The flat organization has several positive attributes:

- Fewer people are involved in making any final decisions
- Operation is less costly than that of a vertical company—there are fewer expensive managers
- Compared to a vertical company, the flat organization is much easier to understand; it is easy to identify the organizations involved in any given decision process.
- Employees' time is optimized; the employees perform multiple jobs. From a management perspective each employee gives more value.
- Employees tend to learn more in a flat organization than in a vertical one, since the flat organization places more functions into one group than does the vertical company.

However, flat organizational structure has negative attributes:

- Fewer people are involved in making any final decisions, therefore less brain power is employed
- There is the risk of spreading too few people across a problem that needs solving or action
- With fewer tracks or paths for promotion, the company may incur greater salary costs to ensure employee satisfaction. Since it cannot give a title to reward great work, it has to give more money.
- Even though employees tend to learn far more in a flat organization than in a vertical one, there is a remote probability that the employees will form their own opinions and act on their own initiative without senior oversight.

Figure 3.7 illustrates the points just made.

The vertical company is closely associated with the classic model of a bureaucracy. It might appear to be focused on "empire building," with many different groups and organizations in charge of various processes or activities. The vertical company can be described as a large number of generals, colonels, majors, captains, and lieutenants. A vertical company

Advantages to a Flat Organization

- Fewer expensive managers
- Much easier organization to understand
- Fewer people involved in making decisions
- People need to learn more

Disadvantages to a Flat Organization

- Fewer people involved in making decisions
- Risk of lacking resources to handle problems
- Run the risk of higher salaries
- Remote possibility that staff will act on personal opinions

FIGURE 3.7 Flat company: advantages and disadvantages.

tends to subdivide tasks into smaller pieces of activity than a flat company would. The vertical company has a number of negative attributes:

- In comparison to a flat company, there are more people in a vertical company involved in making a final decision. Unless the teams are managed properly, there is the risk of introducing too many possibilities for decision-making troubles. In other words, "Too many cooks spoil the broth."
- Employees are given less responsibility and therefore learn less than those in a flat company do.
- The vertical company decision-making process can be difficult to understand since so many different departments are involved in performing any given activity.
- There are many expensive managers or too many mid-range salaried managers. In the end this usually means a huge payroll to support.

We must not forget, though, the positive attributes to the vertical structure:

- The vertical nature of the company enables senior management to create more positions, trading money for titles. In other words, it pays less money as a reward and instead promotes the person into a titled position.
- There is reduced risk of a single group of personnel or a single person working to solve a problem, since more people are involved in a process. The greater division of labor requires senior management to involve more people to get the job done; more people theoretically have more brain power.
- The vertically organized company can more easily focus large groups on smaller-sized tasks.

Figure 3.8 illustrates the positive and negative attributes of a vertical company.

Organizing a company is a challenge. Running the company is another challenge. The division of labor or decision processes must be established so that the company can operate effectively. Years ago, the terms purpose, process, and place were used to describe the areas around which corporate decisions and other processes were organized. Today the terms have changed to product, function, and territory. More terms added to the list of parameters are: time, quality of service, cus-

Negative Attributes to Vertical Company

- More people involved in making a decision – "Too many cooks spoil the broth"
- Employees given less responsibility
- Complicated decision making process
- Many expensive managers

Positive Attributes to Vertical Company

- More career paths in a company
- More resources to throw at solving a problem
- Can focus many resources on many problems, simultaneously

FIGURE 3.8 Vertical company: positive and negative attributes.

tomer segment, equipment type, and alphanumeric organization. These are all parameters of an organizational principle called *departmentation*.

Telecommunications carriers need to be organized so that they can provide service to their customers. It is important to remember that carriers are expected to be available nearly 100 percent of the time, 24 hours per day, 7 days per week. Even though the same cannot be said of Internet service providers, the voice telecom carriers are expected by their customers and regulatory bodies to provide service 99.9999 percent of the time all year long. At some point, the same standard of performance will be demanded of the Internet service provider. Departmentation is the only way for any telecommunications carrier or profit-making enterprise to create the organization needed to provide service. Figure 3.9 is a depiction of the role of departmentation in the telecommunications industry.

DEPARTMENTATION

Departmentation is an organizing principle in which a company creates organizations based on certain parameters or key criteria. The forms of departmentation are:

- *Product.* Large manufacturing companies are typically organized along product lines. An example is the automobile manufacturer. Within the product line, the group may be organized

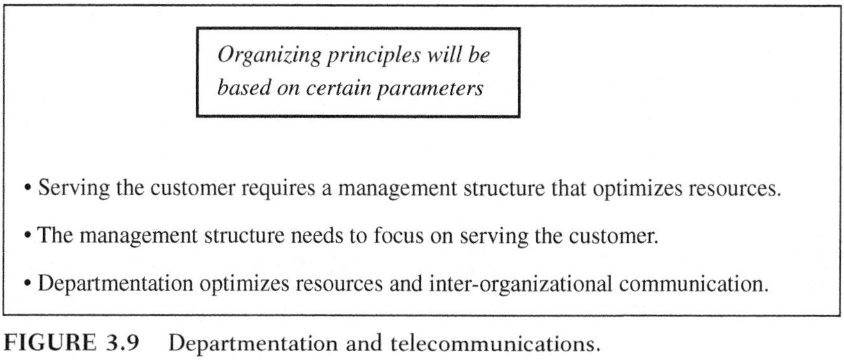

Organizing principles will be based on certain parameters

- Serving the customer requires a management structure that optimizes resources.
- The management structure needs to focus on serving the customer.
- Departmentation optimizes resources and inter-organizational communication.

FIGURE 3.9 Departmentation and telecommunications.

along functions. Product-organized companies like automobile manufacturers are able to create self-contained product organizations. For example, a specific model of car is assembled in a specific plant. All functions required to assemble the car are in one factory. The automobile manufacturers have used product departmentation effectively for decades.

- *Function.* Functional departmentation refers to activities such as engineering, marketing, and finance.
- *Territory.* Territorial departmentation refers to region or divisions of a multinational company that operate as separate companies but still report to a single headquarters.
- *Time.* Time departmentation refers to shift work.
- *Quality of Service.* Quality of Service departmentation refers to classes of service, such as first class and coach on a plane or train.
- *Customer segmentation.* Customer segmentation focuses on specific customer segments. In a telecommunications company, the customers may be broken down into residential or business. A bank may have a loan department focused on farmers and another department focused on small businesses.
- *Equipment type.* Automobile manufacturers may have all drill press work done in a single area, while welding is done in another area.
- *Alphanumeric organization.* Alphanumeric departmentation is common in many companies. In a telecom carrier, customers with last names beginning with the letters A through L are apportioned to go in one direction, while everyone else goes the other way.

Departmentation parameters can be used simultaneously to varying degrees. However, the company is primarily organized around one parameter. Both vertical and flat companies use the same basic principles of departmentation. One of the differences between the vertical and flat companies is in the degree to which various levels of departmentation are applied. Figure 3.10 is an illustration of the forms of departmentation.

FIGURE 3.10 Forms of departmentation.

The most popular forms of departmentation in business, and especially the telecommunications industry, are product departmentation and functional departmentation.

PRODUCT DEPARTMENTATION

As I noted, large manufacturing companies are typically organized along product lines. I gave the example of the automobile manufacturer. Within the product line, the group may be organized along functions. Product-organized companies like automobile manufacturers are able to create self-contained product organizations. For example, specific models of cars are assembled in a specific plants. All functions required to assemble the car are in one factory.

The telecommunications manufacturer will organize its product lines in a similar fashion. Telecom manufacturers will have all hardware work associated with a particular telecommunications switch or router handled by one facility. The software efforts may be handled in another location.

Note that the software groups may even be dedicated to the specific product line.

Telecom service providers will have subgroups within a department organized along product lines. For instance, the marketing department will have a product marketer and a product manager dedicated to a specific type of telecommunications service. However, all marketing work is organized under a single vice president. Figure 3.11 is an illustration of how product departmentation works in a telecommunications carrier and a manufacturing company.

We should remember that in real life, companies are organized around multiple forms of departmentation. However, the company is managed overall by a specific form of departmentation. This concept will be explored later in the book.

FUNCTIONAL DEPARTMENTATION

Functional departmentation is the most widely used and recognized form of departmentation, not only in telecommunica-

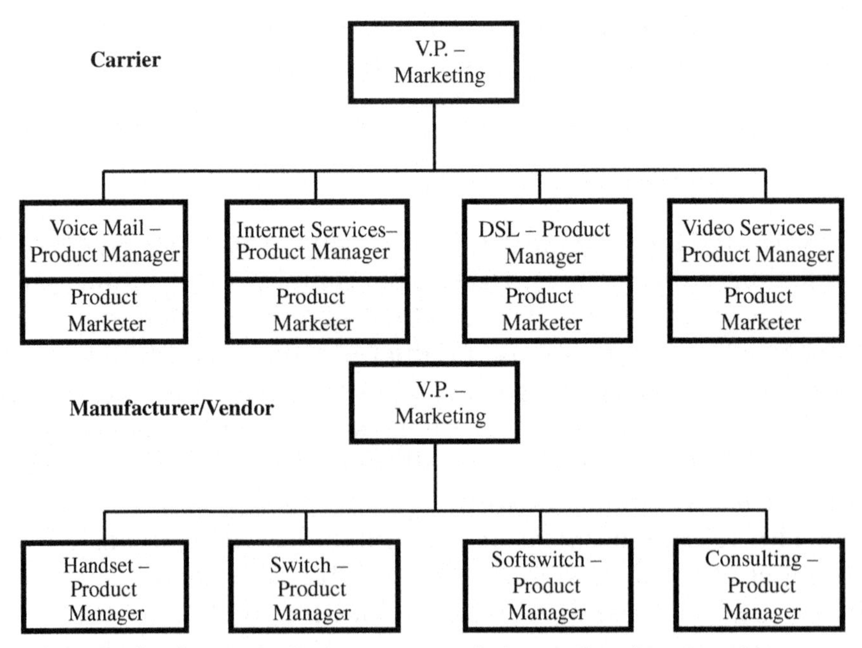

FIGURE 3.11 Product departmentation in a telecommunications carrier.

tions, but in all forms of industry. Functional departmentation is how telecom carriers and large organizations are organized.

One of the goals of organizing a company is to group similar functions under one roof. In other words, the marketing functions are grouped under one departmental roof called Marketing. There are obvious reasons for grouping like functions together. These reason include:

- Economies of scale
- Sharing of resources
- Sharing information—interorganizational politics is no longer a factor
- Decreased response times to events

If we look at the following telecommunications carrier functions, we see how we can group these activities together. The subfunctions can be grouped into functional categories. Each one of the functions shares something in common with at least one other function.

- Sales and Marketing
  - Product marketing
  - Product management
  - Business development
  - Sales
- General Counsel and Government Affairs
  - Regulatory
  - Legislative
  - Legal
- Office of the Chief Technology Officer
  - Technology development
  - Network planning
  - Radio frequency engineering

- Network Operations
  - Technicians—network
  - Network management
  - Installation
- Office of the Chief Financial Officer
  - Carrier access billing—billing between the carrier and other carriers
  - Finance
  - Comptroller
  - Internal Information Technology (IT)
  - Technicians—internal communications
- Office of the Chief Operating Officer
  - Technicians—customer
  - Customer care
  - Customer billing
- Public Relations
  - Media relations
- Human Resources and Administration
  - Building facilities
  - Human resources—regional
  - Human resource—engineering
  - Human resources—operations
  - Benefits
  - Administration—secretarial pool, audio visual, mail room staff
- Office of the President
  - Strategic planning

This is not a complete list of the functions within a carrier. The point is that the carrier groups certain functions into categories. It may not be apparent, but carriers are organized around primary functions that include marketing, sales, customer care, engineering, operations, finance, legislative and regulatory, and legal.

Figure 3.12 is an illustration of these primary functions.

Note that these primary functions usually end up being supervised by senior executive positions, as outlined in the following vice presidential categories:

- General Counsel and Regulatory
- Public Relations
- Chief Technical Officer
- Chief Financial Officer
- Sales and Marketing
- Chief Operating Officer
- Human Resources and Administration

All these positions report directly to the President and CEO. Figure 3.13 is an illustration of the senior management structure. The sections that follow will delve deeper into that structure. To understand how and why functions are organized, we must first understand how and why the senior decision makers are organized the way they are.

- Marketing
- Customer Care
- Sales
- Engineering
- Legal
- Legislative and Regulatory
- Operations
- Finance

FIGURE 3.12 Primary carrier functions.

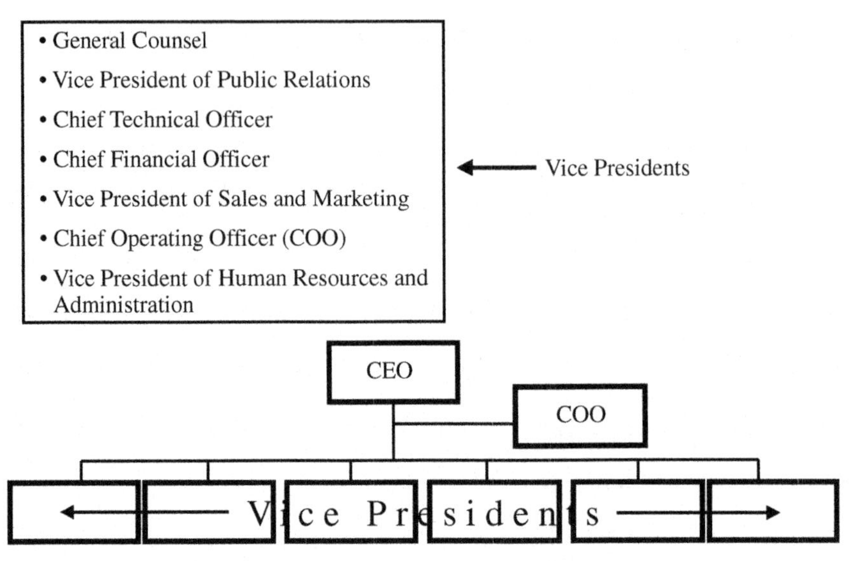

FIGURE 3.13 The Senior Executive organization chart.

BALANCING FINANCIAL OBJECTIVES AND ORGANIZATIONAL NEEDS

Understanding how and why a telecommunications carrier is organized is important from a financial perspective. Carriers are capital-intensive operations. The telecom service provisioning business has always been capital intensive. Constant reinvestment is an attribute of the business. A very good and experienced management team is needed to ensure that the company's assets and people are managed properly. Carriers are like living entities, similar to the people they serve. A carrier needs to be able to respond quickly to the rapidly changing needs and desires of its customers. In order to respond this way, a carrier needs to understand its business and its customers. This balancing that can only be achieved with a good business plan, financing, employees, and a management team.

SENIOR MANAGEMENT

Senior management is a generic term used to refer to the most senior decision makers of the company. The senior management team can be comprised of both the executive team and the senior executive team. The difference between the two lies in monetary authorization levels. The thinking in a large company is that the sheer number of employees requires the creation of more than one layer of executives in order to divide the management responsibilities among more senior managers.

The senior management team in a carrier is divided into the following areas of responsibility: legal, regulatory, legislative, finance, technology, operations, sales and marketing, customer care, human resources, and public relations. Since the functions in a carrier range from installation to billing, each one can be generalized enough to find a place in one of these categories.

These categories are used as a basis for the creation of organizational structures, which take the form of the departments or organizations outlined below. Organization and department are generally interchangeable terms. However, an organization is usually a very large group of people all performing different activities, though these activities share many traits. Therefore a single vice president is in charge of the overall organization. The word department is used by most companies to denote a vice presidential organization. Sometimes a department lies within an organization.

Most management courses teach that there is a theoretical maximum number of people with whom any given person can interface and whom that person can manage on a direct basis. That number is eight. With any more than eight people, the top manager runs the risk of spreading himself too thin and not doing a proper job. This concept applies to any manager-employee relationship, which is why there are levels of reporting. It happens that the number of executives that typically report directly to the office of the company president and chief executive officer is seven:

- General Counsel and Vice President of Legislative/Regulatory—legislative, regulatory, and legal affairs
- Vice President of Public Relations
- Chief Technical Officer—technology and network operations

- Chief Financial Officer
- Vice President of Sales and Marketing
- Chief Operating Officer—customer care and billing
- Vice President of Human Resources and Administration

The following sections describe the activities of these vice presidential positions and departments. Figure 3.14 is an illustration of the vice presidential positions that operate in a carrier.

The need to keep activities of a similar nature grouped together may not be obvious to those unfamiliar with organizational theory. Think of a problem that needs to be solved and think of the activities involved with solving that problem. Those people whose ongoing activities are directly affected are typically in the same functional group. Functional departmentation is focused on grouping activities with similar goals. It facilitates problem solving in a service company and increases a carrier's ability to respond to service troubles.

GENERAL COUNSEL AND VICE PRESIDENT OF LEGISLATIVE/REGULATORY

The General Counsel and Vice President of Regulatory are normally one person in a small or medium-sized carrier. The large

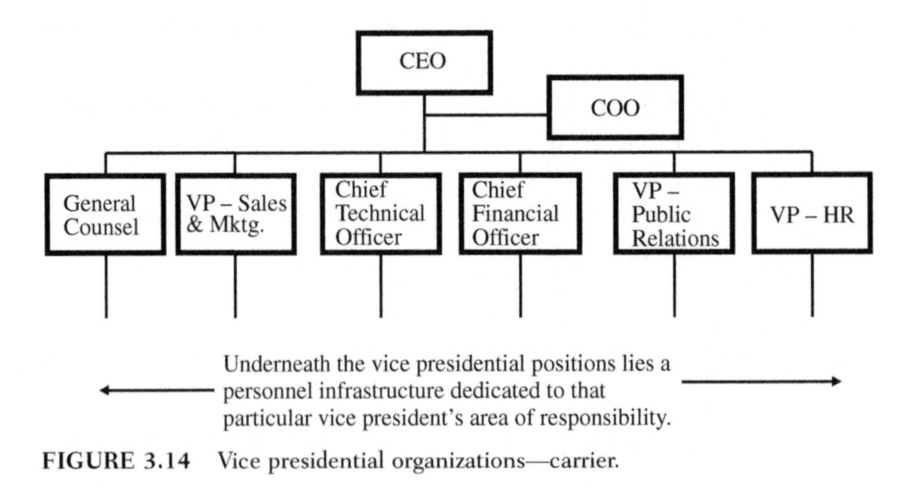

FIGURE 3.14 Vice presidential organizations—carrier.

carriers, such as the old Baby Bells, have legislative (lobbying) and regulatory functions in one department and legal functions in another department. The regulatory and legislative efforts of a carrier are intertwined and require an organization that can manage the functions under one roof. This department is affectionately called "leg-reg."

LEGISLATIVE REGULATORY AREA

The regulatory functions of a carrier are concerned with promoting its regulatory initiatives. The regulatory people work directly with Federal, state, and local regulatory bodies. Leg-reg develops various positions that result in positive regulation decisions. The regulatory area of leg-reg also actively lobbies the various governmental regulatory agencies. It is unlike the regulatory departments in any other types of companies. The regulatory area is one of the most critical and undervalued departments of the telecommunications carrier.

In the telecom service provider business, the regulatory group works to create an environment that mines every penny in a telephone call for the carrier. The regulatory people will find that tenth of a penny in the setup of a call and justify why the carrier is entitled to it. The regulatory group will testify before regulatory commissions, answering questions involving a variety of matters. ILECs are the carriers forced to answer such questions from state/local/Federal regulatory authorities. Figure 3.15 is a graphical depiction of the role of the regulatory area.

The types of questions and issues addressed by the regulatory group within a carrier include:

- Policy
- Costs
- Construction details
- Labor
- Public Obligations

FIGURE 3.15 Regulatory.

- Pricing strategies
- Operating costs
- Cost of provisioning a specific customer service
- Network growth plans
- Construction costs
- Real estate costs for carrier operations
- Equipment costs—Specific detail is always provided as well as overall costs. These costs include tools as well as the large switching equipment.
- Annual state-of-the-business information—operating and customer related
- Quarterly and annual customer trouble reports—aggregated data
- Monthly trouble reports—Specific detail must be provided as well as aggregated statistics.
- Service objective reports—This is data related specifically to key measures of network and overall company performance. The types of data used to measure a wireline telephone company are dial tone delay, number of customer trouble reports in a month, number of customer trouble reports left unresolved in a month, the average time to resolve a customer trouble, types of customer trouble reports, types of network outages, numbers of times switches are taken out of service for maintenance, and maintenance schedules for all telecommunications equipment.
- Employee data—numbers of employees and types of jobs held.
- Customer data—number of customers, number of calls, whether the calls are local or long distance, whether the calls were completed or not.

In an attempt to create broad categories, I have tried to cover every conceivable information type. The point is that the regulatory bodies, especially state regulatory agencies in the United States, require the ILECs to provide details on every aspect of operations and maintain this information for a number of years.

The most heavily monitored telecommunications carrier is the wireline carrier, especially the ILEC. The large wireless carriers, especially those owned by the Baby Bells and the Canadian Bell company, maintain large regulatory groups and report their results to a national agency. In the United States, this agency is the Federal Communications Commission.

In the large wireline and wireless carriers, the regulatory departments maintain close ties to the engineering, sales, and finance departments. A poorly run regulatory department can spell the difference between failure and success for the carrier.

The legislative side of the carrier works very closely with the regulatory side. Its job is to promote government legislation that supports the goals of the carrier. The legislative component of the carrier is a combination of skills and talents. It lobbies for issues, meets government leaders, works to change laws, works to create new telecommunications laws, and emphasize work issues on a state-by-state basis. By the very nature of what the legislative group does, it must be "connected at the hip" to the regulatory group. Hence the reason why the two functions reside in the same department. Figure 3.16 is a representation of the legislative group.

LEGAL GROUP

The legal group is a part of the department known as the Office of the General Counsel and the Vice President of Legislative/Regulatory. It addresses those issues that are legal in nature:

- Lawsuits filed against the carrier by employees, customers, competitors, and government

- Promoting the creation of favorable laws
- Representing the companyís interests in government
- Working closely with the Regulatory Department
- Managing the relationship between the company and all areas of government

FIGURE 3.16 Legislative.

- Lawsuits to file against competitors and vendors
- Working with outside counsel on appropriate issues
- Working with law enforcement to prosecute employees or others who have been arrested committing crimes against the carrier
- Ensuring the company (carrier) complies with local and state laws. The legal department does not oversee compliance; rather it ensures all company departments are informed of what the laws say.
- Ensuring the carrier complies with federal laws
- Writing and executing contracts between itself (the carrier) and the vendor
- Advising the executive staff on a variety of legal issues. Typically, the legal department works to ensure the executive staff makes decisions in a manner that does not violate the laws of the country, state, and municipality.

The legal group does not instruct or dictate; rather the lawyers in the group advise. Figure 3.17 is an illustration of the legal function in a carrier.

By the very nature of what the legal group does, it must be closely integrated with the legislative and regulatory groups. In very large carriers, the legal group is typically an organization in itself. In smaller carriers, the legal function is combined with the legislative and regulatory group.

- Litigation
- State, Federal, and local law compliance
- Contract development

The legal, legislative, and regulatory areas are integrated with one another.

FIGURE 3.17 Legal.

VICE PRESIDENT OF PUBLIC RELATIONS

The Vice President of Public Relations is responsible for the company's public image and the public image of the executive staff. A well-run public relations department can be a powerful company asset. Public relations works with the marketing department and, to some extent, with the leg-reg group to create an image of the company. Advertising is the responsibility of the local territory, with oversight from headquarters. Typically advertising is under the control of public relations with direct input from the sales and marketing departments.

The public relations department works with all forms of external media and information agencies. Public relations is often broken down into specific areas of concern:

- Media relations—television and print
- Industry conferences
- Community action groups
- Public events—e.g., sporting events
- Government—Federal and State
- Local/municipal affairs

Bear in mind that image management is an art, but one that must be executed perfectly the first time around. One flub by an executive could mean a one-point decrease in a publicly held carrier's stock; this might be translated into hundreds of millions of dollars in lost value. Many people consider public relations (PR) as jobs for the mental lightweights, but they are wrong. The PR department serves as the voice of the company and provides spokespeople to the public and all forms of media.

The PR person may not have a degree in nuclear physics, may not understand the intricacies of the corporate balance sheet, or may not understand how the products the company sell work, but that is just fine. The PR person is as critical to a company as the telecom finance person. Figure 3.18 illustrates the general role of the public relations person.

- The Public Relations department is the face of the company.
- Engages the news media.
- Manages the company's public image.
- Manages the management team's interactions with the media.

FIGURE 3.18 The role of public relations.

A PR person is needed even in a startup. At some point in a startup company's life, it must announce its presence to the world. The PR person has the relationships in the media world to make sure the company gets the column space in the "trades" (trade journals) and the mainstream media to get the word out about the company. Figure 3.19 illustrates public relations' role in a startup.

EXAMPLE OF PUBLIC RELATIONS

Given its perceived fluffy and amorphous nature, the public relations side may need a little more explanation. PR is too important to ignore or take for granted.

Creativity in publicizing the company

Creating industry excitement

Creating interest in the investment community

FIGURE 3.19 PR and the startup carrier.

PR is not just required to promote a company's image. It is needed to mitigate the troubles that arise from a telecom network disaster. For example, some years ago a large U.S. carrier suffered a major SS7 network failure. Several states in the Untied States were affected, and millions of customers were unable to make or receive telephone calls. The public relations department was immediately dispatched to handle the ensuing cries for justice against the telephone company. The Federal government demanded an investigation, community groups cried out for help, all insisting some form of punishment be leveled against the telephone company; individuals began to file lawsuits; and newspapers printed all the stories.

The PR department in the telecom carrier, with the other telephone company departments, worked to create statements for the executives and other key personnel that would best explain the network failure and how it occurred. In the midst of the SS7 network disaster, the PR team was speaking to every television news program, newspaper, magazine, and trade journal in the U.S. Many telephone company employees stopped thinking these people were just talking heads. Without PR, the executives would not have known how to make the appropriate statements without offending the public or even leading to additional lawsuits. The PR people did an extraordinary job of maintaining the image of the company and keeping the company's stock from being more adversely affected than it was.

PUBLIC RELATIONS—PERCEPTION IS REALITY

All advertising has input from the public relations department. It must sell the product and the company. When a representative of the company is speaking at a public engagement, everything from logo use to the types of questions that cannot be answered are handled by public relations. The PR department does not tell the speaker what questions to answer, rather it works to ensure that the individual does not say the wrong thing. For instance, an employee making an off-the-cuff comment about the mobile handset selection process would be advised to say something like "that is not in my area of expertise or responsibility." I once saw an employee of a large carrier talking about the future of handsets in the wireless market (at a large industry conference) answer a question about his company's process of selecting handsets for customers. His response was "the company just selects the cheapest handset it can find, marks up the price, and then sells the handset to the customer." To some, it sounded as if the engineer were trying to be amusing; to others, he sounded as though the question were a waste of his time. I do not know what happened to the employee, but I did see looks of shock on the faces of people who were from his company and there were many people in the audience laughing. I took the comment as a speaker's joke. However, it was obvious from his appearance, later on, that his employers did not take it lightly. Of course, it did not help that conference attendees were constantly re-telling the comment and that the trade press was at the conference looking for a story.

Then there was the time a president of a major carrier told an off-color and rude joke to get the attention of conference attendees. This incident occurred at a time when the industry segment was in its infancy; there must have been only 50 people in the audience, and everyone knew this person. We all knew we had been rude to him by not paying attention to him and shouting and whooping at him when he ascended the stage. We deserved the snipe. We all laughed and paid close attention. His joke was totally inappropriate and today, we are all embarrassed that we laughed. Years later, a colleague of mine tried to tell a similar joke in front of audience of about 500. This time the joke failed and an apology was made. The audience was offended and people demanded an explanation of how and why this person was allowed to speak. The word on everyone's lips at the conference was that the company my colleague represented was comprised of unprofessional telecom people. Some competitors even went as far as saying the individual was an example of the type of people the company hired and the work they represented. The timing of comments, rumor, and harsh feelings all have a way of wrecking a company's future and image. PR people are needed to keep employees in check and under control.

Many telecom executives travel with a least one PR person in tow. The job of the PR person in such a situation is to make sure the executive sees and speaks to all of the right media representatives.

In the real world, perception is reality. The public relations department's job is to create a positive public perception of the company. The public relations team cannot make things up out of thin air, but a good team can come fairly close to turning sand into water and wood into steel.

CHIEF TECHNICAL OFFICER—TECHNOLOGY AND OPERATIONS

The Chief Technical Officer is responsible for all technology planning, engineering design, and network operations issues in the company. The title has gone through a number of transformations in the last 50 years. The CTO was called the Chief Engineer in the old wireline telephone companies. The position was then called Vice President of Engineering and Operations. Next, it was called Chief Technology Officer. Now the popular term is Chief Technical Officer.

In some carriers, the position of CTO denotes someone who is responsible only for new technology planning and deployment management. In other carriers, the Chief Technical Officer is responsible for technology planning, deployment, and day-to-day operations. In still other carriers, the Chief Technical Officer is responsible only for technology planning.

The reason this position is given such a lofty title is that the individual holding the job is responsible for spending nearly 70 percent of the company's annual budget. The network is the asset of facilities-based carriers. The network requires large amounts of money to build, maintain, and operate. The Chief Technical Officer's technology planning responsibility involves the following tasks:

- Current planning—capacity planning needs for the next two years
- Usage forecasting—this requires direct input from sales, customer care, and marketing

- Evaluating new and existing technology deployment during the next two years
- Evaluating how technology is deployed during the current planning period
- Evaluating the type of technology to be deployed during the current planning period
- Evaluating new technology deployment for the period of three years in the future and beyond
- Evaluating how technology is deployed during the period of three years in the future and beyond
- Evaluating vendors of technology that has not even been beta tested
- Planning all aspects of technology implementation—establishing schedules for installation and testing, establishing all implementation procedures, and establishing all testing procedures
- Overseeing and managing the day-to-day operations of all areas of the network, excluding customer support
- Operating and managing the network control centers
- Operating and managing the transmission facilities network
- Evaluating and selecting all telecom equipment vendors
- Responsibility for all telecom equipment vendor contracts
- Close coordination with sales and marketing

Figure 3.20 illustrates the responsibilities of the Chief Technology Officer's organization.

Given the number of tasks this department manages and the disparate number and types of skills needed, typically departments are formed within the technology organization. These include:

- Network planning and engineering
- Network operations
- Strategic network planning

- Evaluating technology
- Evaluating vendors
- Selecting technology
- Installing
- Maintenance
- Operations
- Design

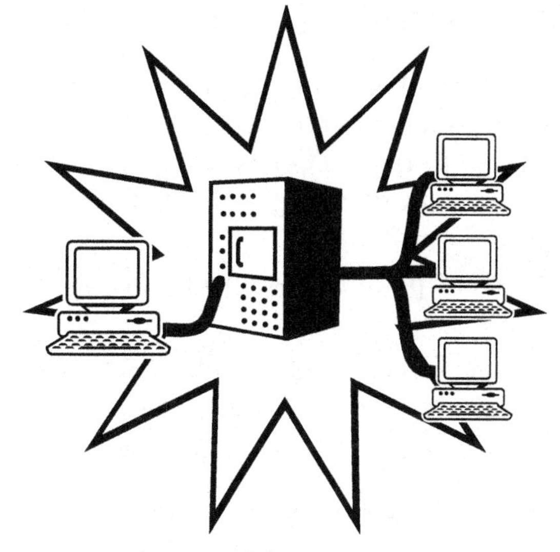

FIGURE 3.20 Technology organization's responsibilities.

Figure 3.21 is an illustration of the departments within the Technology Organization.

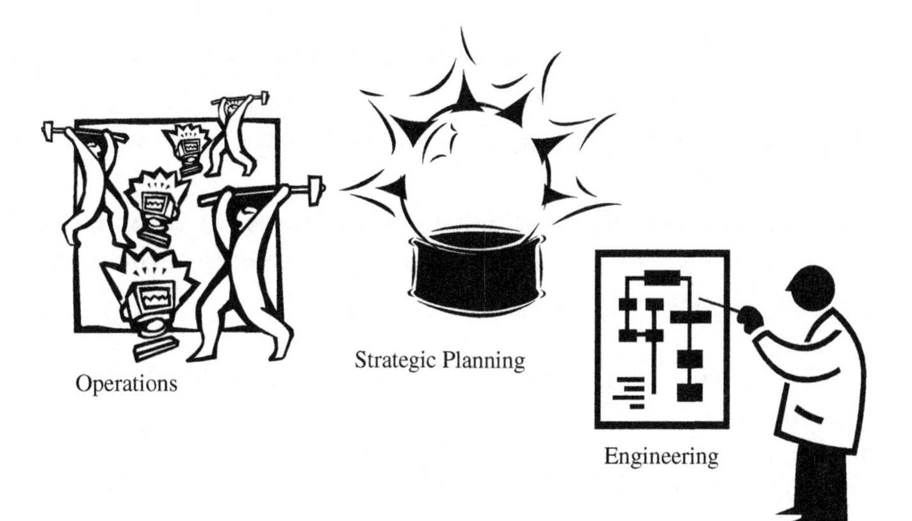

FIGURE 3.21 Technology departments.

NETWORK PLANNING AND ENGINEERING

The Network Planning and Engineering Department is run by a vice president who reports to the CTO (Chief Technical Officer). Effectively, the CTO is a senior vice president. The activities of the Network Planning and Engineering Department are:

- Current planning—capacity planning needs for the next two years
- Usage forecasting, which requires direct input from sales, customer care, and marketing.
- Evaluating new and existing technology deployment during the next two years
- Evaluating how technology is deployed during the current planning period of the next two years
- Evaluating the type of technology to be deployed during the current planning period
- Planning all aspects of technology implementation—establishing schedules for installation and testing, establishing all implementation procedures, and establishing all testing procedures
- Evaluating and selecting all telecom equipment vendors
- Responsibility for all telecom equipment vendor contracts
- Close coordination with sales and marketing to ensure that installed equipment is capable of providing the services being sold to consumers
- Usually, responsibility for writing requisitions for equipment; sometimes this responsibility falls into the hands of the operations department

This particular department has control of the largest budget in the company because it is responsible for the construction of the network. In the old Bell Telephone companies, this organization was called Network Engineering.

It should be noted that usage forecasting is a function that resides across multiple organizations in a carrier. The technology organization cannot forecast consumer patterns without input from the organizations that actually interact with customers. Customer forecasts are usually created by Network Engineering or Marketing.

STRATEGIC NETWORK PLANNING

The Strategic Network Planning Department is run by a vice president who reports to the CTO. The department is responsible for:

- Evaluating new technology deployment for the period three years in the future and beyond
- Evaluating how technology is deployed during the period three years in the future and beyond
- Evaluating vendors of technology that has not even been beta tested

You may wonder how the three-year period was established: It takes about a year to deploy a large network's worth of switching or routing equipment. In year two, the carrier is aggressively selling and marketing services associated with the new technology. During the second and third years, customers are using the service and the carrier is gathering and processing consumer data. By the time the third year has come around, the consumer is looking for something new. The tax structure of the country has an impact as well. In the United States, a great deal of telecommunications equipment is amortized over a three- to five-year period. Switching equipment is amortized over a seven-year period for now. Therefore, the first three years is the current planning period.

The responsibilities of the Strategic Network Planning Department do not have as immediate an impact on the technology of the company as the responsibilities of the Network Planning and Engineering Department do. However, the department's work can affect the future direction of the company.

For example, say a carrier whose primary product is voice is looking at deploying a new switch capable of processing both voice calls and data transactions. The switch is, in fact, a combined time division multiplexed switch and an Internet router. This sounds like a good product for a carrier to purchase and install, especially since the carrier wants to enter the Internet business space.

However, the Strategic Network Planning Department has been reviewing the development work of a manufacturing company who has traditionally sold to Internet service providers and is attempting to enter the voice business. The Internet equipment manufacturer is exactly six months away from having a sales-ready product and wants to give the carrier "most favored nation" status. This is an industry term to describe highly favorable terms given to a carrier by a manufacturer. The reasons for awarding this status can be numerous, but, typically, the largest carrier gets the status so that the manufacturer gets the "big contract" and makes lots of money. Carriers often say that a vendor will do anything for a very large customer.

The Strategic Network Planning Department usually focuses on value-added technologies, those that can be deployed as overlays to the main switching/processing matrix (network). Changing out switches or routers is an expensive activity.

NETWORK OPERATIONS

The Network Operations Department is run by a vice president who reports to the CTO. Its responsibilities are:

- Overseeing and managing the day-to-day operations of all areas of the network, excluding customer support
- Operating and managing the network control centers
- Operating and managing the transmission facilities network

The Network Operations Department can be a completely separate organization having its own senior vice president, one not reporting to a CTO. However, it may not be wise to separate operation functions from technology functions. The activities are too closely related to be in two different organizations. A single overall organization head can ensure that the company's network goals are achieved without organizational fighting. It is not impossible to separate the departments, but the separation requires too much management coordination. Reconciling the personal agendas of two independent organizational leaders takes a lot of time and energy. When one person is in charge of the overall organizations, there is the advantage of having both a referee and a final decision maker. Figure 3.22 is an illustration of the role of Network Operations.

The Network Operations Department is responsible for the following discrete activities, whose subsections are summarized in the discussion that follows. The summaries are applicable to the management of any network:

- Network element management
- Network systems management
- Service management

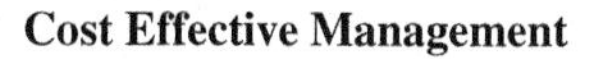

- Network monitoring
- Network health
- Management
- Repair
- Metrics

FIGURE 3.22 Network operations.

NETWORK ELEMENT MANAGEMENT. A network element can be defined as a database, switch, router, voice mail recording system, or adjunct system. Network Operations oversees a number of activities and functions, including:

- *Maintenance schedules*
- *Routing table update schedules*
- *The number of subscribers being served by the system.* Without knowing how many customers are being served, the network manager cannot properly manage the traffic load of the network. Traffic load management entails moving telecommunications traffic from one part of the network to another and even blocking new traffic to ensure that existing traffic is properly routed.
- *Services being provided to the subscribers.* Network managers need a complete picture of the services available to the subscriber base. These managers are not concerned about who specifically has what subscriber service, but about the services available to the population of subscribers as a whole. In some instances, specific hardware or software may need to be available in the network in order to provide a service.
- *Performance objectives for the network element.* Measuring the performance of a network element is way of measuring network health. The network manager needs to have an understanding of how each element in the network is expected to behave. Using these benchmarks, the network manager can address malfunctions or anomalous conditions.
- *Transaction processing time thresholds.* The processing time of any transaction is another indicator of network health.
- *Failure levels/indications.* Any network element will fail at some point in the life of the device. The failure needs to be reported to the network manager. The network operations centers need to be able to reflect the failures in some manner, by audible or visible alarms.
- *Alarm condition indications.* Anomalous conditions also need to be announced in a visible or audible manner. More

important, the alarm indication should assist the network manager in resolving the alarm condition. This means the indication must report a meaningful event; within the context of network management, the event must be real network trouble or an unusual occurrence.

NETWORK SYSTEMS MANAGEMENT. Network systems management involves the way the entire collection of elements and all the functions within the business that affect the physical network are managed. Network operations managers seek to address the same questions and issues about the whole network that are addressed for the individual network element. These questions/issues of concern are:

- Service objectives for the system
- Network monitoring software tools
- Network diagnostic tools
- Anticipated traffic load on the system
- The overall traffic profile of the system on a hourly and daily basis
- Types of subscribers being served
- The level of network system control—can the entire system be re-booted or is an element-by-element process involved?
- The possibility of rerouting traffic
- Types of alarm indications
- Types of network failure indications
- Overall network operating procedures documentation
- Disaster recovery plans

SERVICE MANAGEMENT. Service management links the network systems management function with the customer care function, including performance and customer satisfaction metrics. Service management requires involvement with the customer. It involves every mechanism and process needed for delivery of services to the customer: operational support systems, sales, and marketing. This group works very closely with sales, marketing, and customer care.

Service management links the network to the financial portion of the telecommunications service business and addresses the following:

- *Establishing and following service level agreements*
- *Reviewing market and customer satisfaction surveys*
- *Reviewing competitive analyses.* Issues that arise from these analyses are often incorporated into the day-to-day activities of the network operations department.
- *Conducting operations cost analyses on a constant basis.* Operations analyses of this type address the cost of running the network. Often understanding how a competitor runs its network will provide important information on the shortcomings of the competitor's network. Such information can be used to drive up the costs of the competitor's network. Rising operations costs can mean the difference between meeting financial margins and operating in the "red ink" column.
- *Customer complaints resolution.* Customer complaints are processed from the customer care department and are usually addressed by the network operations department. Not all issues are network related: A customer will call a carrier and complain about nearly everything from how expensive a service is to how confusing the bill is.
- *Management of operations cost.* Managing costs is essential to every business. In the case of capital-intensive service providers (carriers that own and operate their own equipment), it is imperative that every minute of the day, every item of spare parts inventory, every procedural step, and every tool, be managed in the most cost-effective manner possible.

CHIEF FINANCIAL OFFICER

The Chief Financial Officer (CFO) is not just the accountant or bean counter for the company, but is responsible for ensuring the enforcement and establishment of fiscal policy within the company. The CFO must:

- Work with the executive team and, possibly, investors, to build a business model that will generate positive revenue
- Establish the company's cost controls
- Establish the company's letters of credit
- Incorporate the company
- File the appropriate security trading commission's quarterly reports and annual reports if the company is traded on the stock market
- Establish vendor payment schedules
- Analyze all contracts to ensure that the company's financial goals are met
- Oversee employee expense policies
- Approve all equipment purchasing and leasing agreements between the company and other parties; these non-telecom equipment agreements include building leases for office space or switching facilities, utility bills, and even office supplies
- Work with investors and company management to establish the financial benchmarks against which the company will be measured; these benchmarks can be the price-earnings ratio or even EBITDA (earnings before interest, taxes, depreciation, and amortization)
- Pay out company taxes
- Manage the carrier's accounts payable and accounts receivable
- Work with the various executives to establish department budgets
- Manage the company's stock options plan
- Manage the company's relationship with stock analysts
- Develop financial models to predict changes in the company's financial health with every change in the business model
- Track the company's overall financial health

- Oversee the company's IT (information technology) department
- Sometimes, oversee the mailroom, which is responsible for accepting external mail and distributing it to employees
- Monitor company revenue generation

The CFO operates across the breadth of the company and is responsible for managing the company's money. In a startup company, the CFO often has the further responsibility of seeking out additional sources of investment dollars. Figure 3.23 is a representation of the Chief Financial Officer's organization.

The CFO is responsible for the company's corporate intranet, the key internal communication system that serves a mission-critical function in the company. It may not be obvious, but a corporate intranet is a company's internal asset. As a company asset, the IT network falls within the jurisdiction of

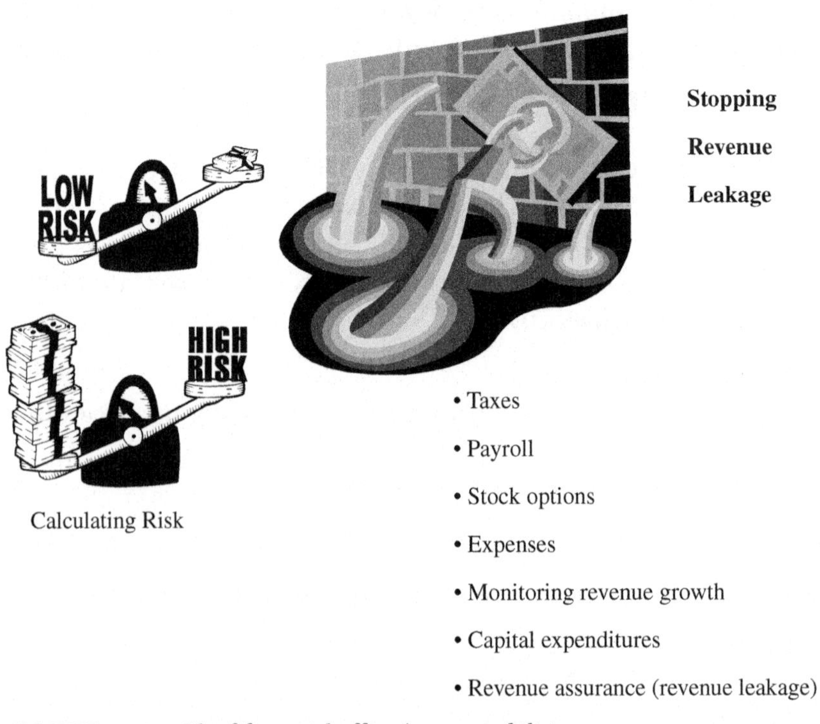

FIGURE 3.23 Chief financial officer's responsibilities.

the CFO. Remember, the CFO manages the IT department as well as the company's leases.

The CFO is typically an accountant or a finance person. I have seen cases in which the CFO was not an accountant or even a finance person but because of the person's strong influence in the financial community, the individual was made CFO. The CFO's responsibilities are broad and can be overwhelming. In a large carrier, the CFO will create three positions:

- Vice President of Information Technology (IT)—This position is a separate executive level position in very large companies; think the big Fortune 100 companies.
- Vice President of Finance/Comptroller

Figure 3.24 is an illustration of the departments within the financial organization.

VICE PRESIDENT OF INFORMATION TECHNOLOGY (IT)

The Vice President of Information Technology (IT) is responsible for managing the company's intranet, its internal computer information system. The IT department manages the various

- Managing corporate intranet
- Financial analysis
- Managing internal costs

FIGURE 3.24 Financial organization departments.

network elements that enable the company's email, data archives, and corporate firewalls. This department used to be called the MIS (Management Information System) department. The IT department replaced the company's internal mail system. Not long ago, large corporations still had mailrooms that managed not only external but also internal mail.

Many readers may remember the inbox and outbox, major fixtures in every office and on every desk. The inbox held all incoming mail to the employee. Incoming mail included mail from external sources and internal mail from other employees. The outbox held all the mail going to people outside the company and mail going to other employees. All mail was picked up and delivered by people who worked in the mailroom. In a company of 100,000 employees, thousands of pieces of mail were moved through the company every day.

In the twenty-first century, the mailroom is replaced by the corporate intranet, which has made it easier for employees to send messages inside and outside the company. Thousands of emails are sent across the corporate intranet every hour. The corporate intranet handles both serious company business and personal jokes. It is a major component of a carrier's day-to-day operation. The old paper internal mail room could take a full day to deliver a document; with the intranet, delivery takes only seconds.

It is impossible to discount or minimize the importance of email service to the Internet and corporate intranet. Email is a mission-critical component of the business. It has assumed an identity that cannot easily be equated to that telephone or cellular communication. Email is the only service to date where a simple communiqué between family and friends may be nothing more than a joke or poem. Email in a company has become more important than the internal telephone call. It has enabled companies to communicate with employees almost around the clock. Email has added a whole new dimension to communication between parties.

Email literally allows one to carry on a conversation in near-real time. Email letters permit people to engage in quick discussions about important and frivolous topics. Copies of letters and other documents can be transmitted with an email. Email has reduced communications latency between parties.

Email has even allowed the carrier to communicate with vendors, who send important documentation to the telecommunications carrier. The network operating statistics of the company can be transmitted by email to remote company locations for review by network staff.

In addition to managing the corporate intranet, the IT department is responsible for managing the slew of desktop and laptop computers used by employees. Managing the computer equipment assets includes maintenance, repairs, upgrading hardware, and upgrading software.

The carrier's IT department may also be responsible for maintaining (and in some carriers managing) all customer-related and day-to-day operational systems.

CHIEF INFORMATION OFFICER. In very large companies, the position of Vice President of IT is distinguished from the finance organization as a separate executive-level position. The companies in question are so large that the IT asset of the carrier requires a focused budget view. The only way to accomplish that is to give the VP of IT direct reporting stature to the CEO.

Some may feel that the CIO's job is not a glory job but a strange job, especially since the carrier is in the information business. The CIO function became popular only in the mid-1980s. The position was first called Vice President of Telecommunications, then became Vice President of Management Information Systems (MIS). Email and internal information processing systems have become so important that the CIO function is the norm and not the exception in large companies. Internal information systems are every non-telecom-related system used in the company: customer care and provisioning systems, billing systems, and any other information processing system. The CIO is responsible for maintaining and managing these systems, but the processes those systems support are the responsibilities of other senior executives.

Small companies have smaller budgets, and the staff needed to manage the IT functions can be more easily managed in an organization that accommodates both finance and information systems.

VICE PRESIDENT OF FINANCE/COMPTROLLER

In small companies, the Vice President of Finance and Comptroller are the same person; he is responsible for the financial transactions between the carrier and the customer and all internal financial mechanisms.

In much larger carriers, the functions of customer accounts receivable and refunds/credits are managed by a vice presidential organization, while another vice presidential organization manages all internal financial mechanisms. In large companies where these functions are divided into two different departments, both are placed under the management control of a senior vice president or a Chief Financial Officer. The reason for the division of labor is that very large customer pools will usually bring with them very large issues. Before we delve into this aspect more deeply, we should take the opportunity to describe all aspects of the job of Vice President of Finance/Comptroller.

The finance person and comptroller are responsible for a number of different tasks:

- Payroll
- Payroll tax processing
- Time sheet processing
- Accounts payable to pay suppliers, landlords, leasing companies, office suppliers, and utilities
- Local, state, and Federal taxes—income, surcharges, and other government fees
- Accounts receivable—customer payments
- Customer credits
- Internal corporate financial practices
- Employee expense reports
- Company departmental budgets
- Operating cash

- Charitable donations from the company
- Government fines
- General accounting, which addresses the planning, developing, implementing, maintaining, and upgrading of internal accounting systems that are used to ensure the carrier's compliance with appropriate telecommunications regulations, policies, and laws that ensure proper revenue accountability.
- Internal financial control mechanisms—creation and enforcement
- Periodic reporting to investors (if the carrier is public) done monthly and quarterly. A public company is normally required by its national regulatory agencies to report its financial health on a periodic basis. If the company is private and large enough, it must still maintain financial accountability to local regulatory agencies. When the carrier is private, determining whether or not it appears on the monitoring screen of a regulatory authority depends on whether that carrier has a significant market position.

The function of the Vice President of Finance is to manage all internal financial mechanisms and internal budgets, while the job of the Comptroller is to oversee all transactions between the carrier and the customer.

The CFO is required to be the gatekeeper of all financial reporting to various government agencies. Unlike many other North American industries, the telecom industry is perceived to be large enough to warrant special attention by Mexican, United States, and Canadian telecommunications commissions. CFOs who have never worked in telecommunications may find the government oversight unnerving. However, telecom service provisioning to the public is not considered a privilege for those who can afford it, but rather a public right for all people. This is a difficult concept for many CFOs to understand, since it is totally political in nature. Figure 3.25 recaps the entire CFO's organization chart.

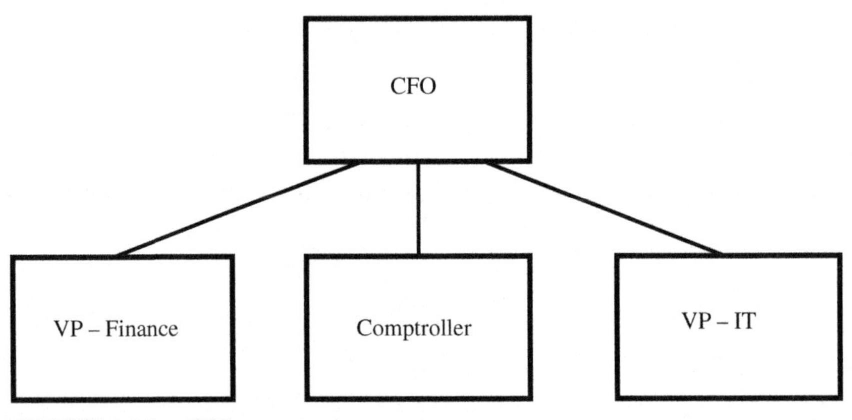

FIGURE 3.25 CFO organization.

VICE PRESIDENT OF SALES AND MARKETING

The one thing any reader of business literature has probably learned is that the company is always selling and marketing its services and products. Those who think they are not selling are mistaken. All companies sell, even the ILECs and the massive long distance and international carriers.

The Vice President of Sales and Marketing in a telecommunications carrier is responsible for identifying customer segments, determining what appeals to those customers, and getting the customers to buy. The job is far more complicated than what I have just described. The head of sales and marketing must:

- Identify market environment
- Define market size and potential
- Describe applicable customer demographics
- Identify potential competitors
- Develop competitive strategies
- Development business
- Define customer segments
- Define customer needs

- Manage product, translating customer needs into understandable requirements for the carrier's technology department
- Define the price points for the product
- Manage all agent sales programs
- Manage all major customer accounts, both wholesale and retail
- Define sales strategies
- Make the sales

This has been a summary of major tasks. Chapter 5 will delve into the area of sales and marketing in greater detail. Figure 3.26 illustrates the area of responsibilities for the sales and marketing organization.

The Vice President of Sales and Marketing divides his area of retail responsibility into two major categories: residential and business. These categories are divided even further into the following groupings:

Market Segmentation

Defining the Market

Defining Customer Needs

Targeting the Market/Customer

Casting Out the Net for Customers

FIGURE 3.26 Sales and marketing areas of responsibility.

- *Residential.* Single family dwelling
- *Residential.* Multi-unit/family dwelling
- *Business.* Small to medium
- *Business.* Large

Wholesale and resale components of the business exist as well. From the perspective of a carrier, wholesale means that the carrier allows a specific company (Company B) to take the carrier's (Carrier A) services and resell those services under the Company B name. Wireless and even wireline carriers have been known to do this. Imagine that a carrier wishes to provide a corporate park's primary telecommunications services. In order to make this deal occur, the carrier will allow resale of its services under the landlord's own name. Many corporate landlords have registered themselves as non-facilities-based sellers of telecommunications services. The landlord does not own any equipment to provide services; he simply sells to his tenants those services Carrier A would have sold to the tenants directly. The reason Carrier A must submit itself to such treatment is that the law states that the landlord is required to provide non-discriminatory access to the tenants; non-discriminatory does not specify who gets into the building first. Further, what better salesperson for Carrier A but the landlord?

From the perspective of the carrier, the resale component of the business is when the carrier is reselling some other carrier's service. In the United States, this has entailed the ILECs' reselling the long distance service of the long distance carrier. The CLEC is either selling its own services or reselling the services of the ILEC. Figure 3.27 is an illustration of an aspect of the services side of the carrier.

A good portion of the rest of this book will be dedicated to the sales and marketing aspects of telecommunications services.

CHIEF OPERATING OFFICER (COO)

The Chief Operating Officer (COO) is usually seen as the number-two person in the company, below the President and

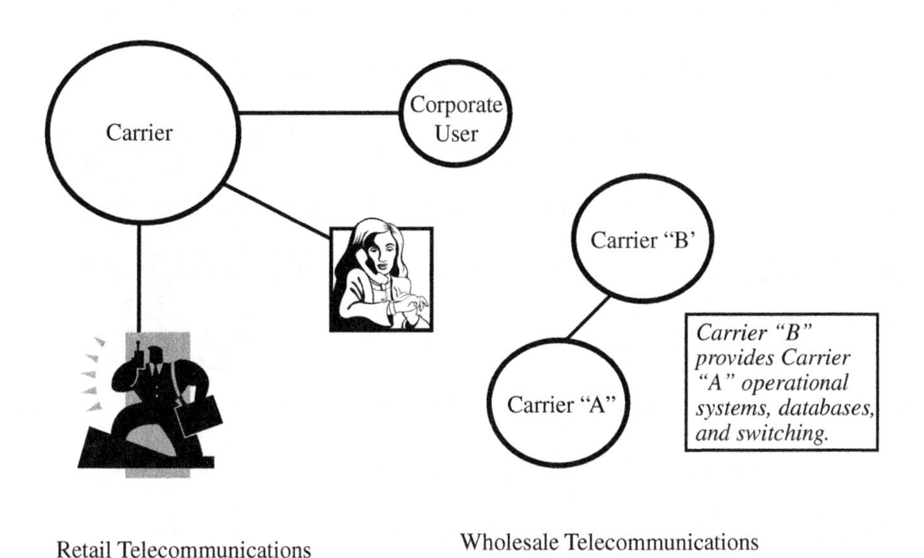

FIGURE 3.27 Retail and wholesale services.

Chief Executive Officer (CEO). What does that mean? The CEO is the one who makes the final decisions. The CEO has a personal staff to sort through information and paperwork. The CEO's staff supports the CEO. So what does the COO do?

In every company, the COO is next in the management line after the president. In some companies, the president is also the Chief Operating Officer (COO). The president in some companies is considered the top person in charge of the day-to-day and tactical decisions. The CEO is considered the final decision-making authority on all company strategy decisions and financial issues. The company's Chairman of the Board is usually the person who faces the stockholders and the private investors. The chairman is the top person in charge. The COO is typically assigned some task that has a customer or financial component, such as customer care. In a carrier, the COO is in charge of customer care, provisioning, and the company's customer billing systems (not billing processes but the physical systems). Figure 3.28 is a depiction of the COO's role in the telecommunications carrier.

This book takes the position that the president of the company is also the CEO. The COO typically fills in for the CEO

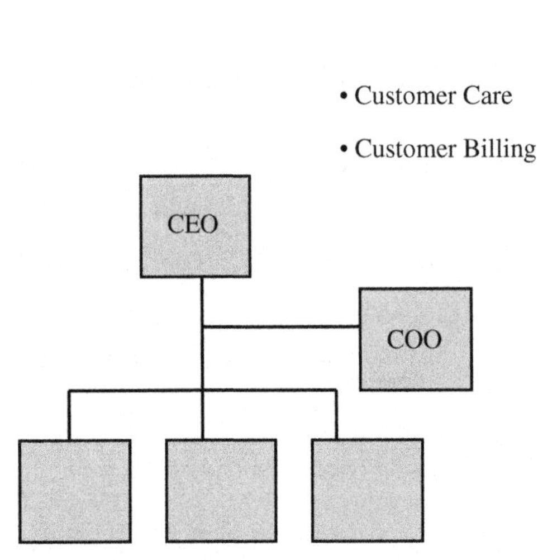

FIGURE 3.28 The COO in the telecommunications carrier.

when the CEO is on vacation or takes a day off. In the case of a carrier, or, for that matter, any company, ideally the CEO and the COO are never out of the office at the same time. In the 1970s, a large U.S. company lost nearly its entire top management team in a plane accident. The company had to operate without many of its most senior managers. It survived, but things were shaky at first. The policy for most carriers is that the COO and the CEO never travel out of town at the same time.

The COO is in charge of the customer care function because this is the area in which the carrier can immediately perceive changes in the customer base and marketplace. The provisioning process is the process of associating services to the customer and the creation of a place within the billing process. The financial role is critical to the carrier, but the customer care function is the bloodline of the company. Customer care is the area of the company that enables the carrier to get an immediate read on customer satisfaction. Unhappy customers mean no revenue. Figure 3.29 is a representation of the importance of the customer.

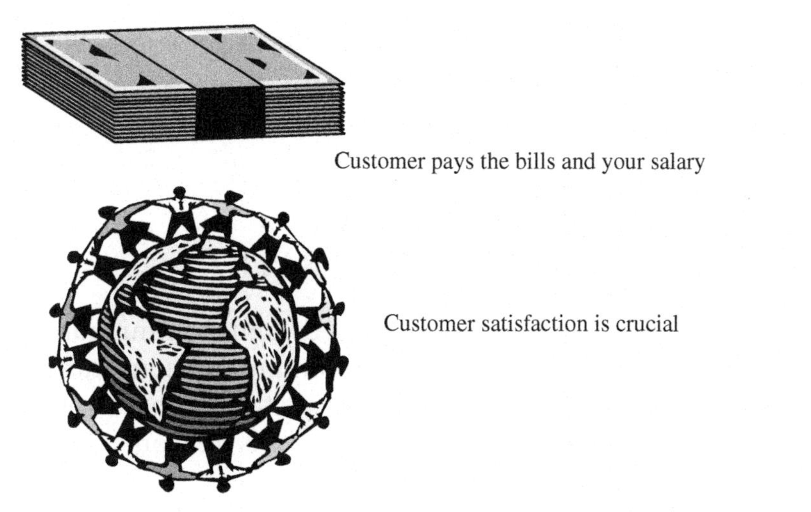

FIGURE 3.29 The customer.

In regard to the billing systems, the COO is in charge of the proper function of customer detail recording and back-office interface systems. The systems are integrated into the switch or router and the carrier's rating systems. The rating systems are those systems that apply the various discounts and rate plans to the customer's bill and are under the control of the CFO. Figure 3.30 is an illustration of how the COO's systems responsibilities are related to those of the CIO and the CFO.

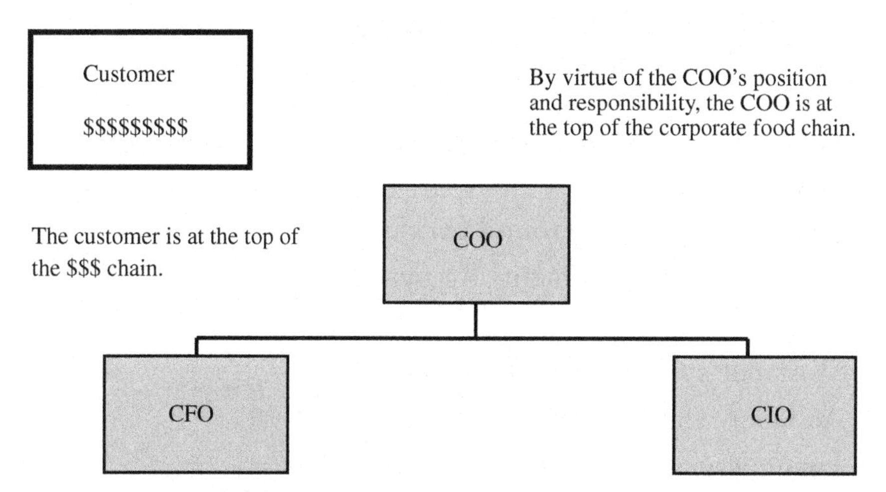

FIGURE 3.30 The COO versus the CIO and the CFO.

In most companies, the physical maintenance of systems is under the control of the CIO, while the processes are the responsibilities of the other senior executives.

VICE PRESIDENT OF HUMAN RESOURCES AND ADMINISTRATION

The Vice President of Human Resources and Administration has the most thankless job in the telecommunications carrier and normally in most other industries as well. The Human Resources and Administration department is responsible for managing the following areas:

- Building facilities
- Human resources
- Employee benefits
- Employee testing
- Hiring and termination practices
- Retirement benefits
- Pensions (if any)
- Parenting leave programs
- Interviews of candidates for positions
- Maternity leave programs
- Company security (in large companies, security is managed by a separate department that reports directly to the president and CEO)
- New employee background checks
- Immigration requirements for new employees
- Employee drug testing
- Workmen's Compensation issues
- All labor relations issues
- Union issues (if any)
- Substance abuse rehabilitation

- Employee occupational and safety guidelines enforcement
- Affirmative action policy enforcement
- Employee code of conduct
- Employee identification cards
- Administration—secretarial pool, audio-visual, mail room staff

The Vice President of Human Resources and Administration (HRA) reports directly to the President and CEO and is often the company's barometer for employee attitude. The HRA department is an organization that goes largely unnoticed by most employees unless they have a particular problem or a company-wide issue arises. The HRA department is constantly operating below the "radar screen" of most employees and that is the way it should be run.

The Human Resources and Administration department's wide span of responsibility ensures that both the corporate officers' and the employees' interests and needs are met. The Vice President of HRA has responsibilities to the carrier's executive team that require him to work closely with the Vice President of Public Relations. Any public statement made by an executive is potentially the fuel for an employee, customer, or supplier lawsuit, or government action. The HRA and Public Relations departments work diligently to protect the executives' images and careers. In addition to ensuring the well-being of the executive team, the departments also work to protect the image and health of the carrier. Carriers sell services and image is everything.

The administrative part of the job makes sure things like office supplies and associated contracts are in place. A carrier with its own cafeteria will have the HRA department administer the cafeteria contract. The meals served in the cafeteria are cooked and provided by a food services company.

These issues may seem unimportant in the business of the company, but they are not. Have you ever had your photocopier run out of toner and found there were no spare cartridges? Have you ever had the photocopier break down? Have you ever had a toilet break in a restroom? Have you ever had the office air conditioning system break so that no one could adjust the temperature? Have you ever needed new doors or shelving installed in an office? Have you ever had an office theft? Have you ever had an office broken into? Have you ever had vehicles stolen from the company's parking lot? The answer probably is yes to at least two of these questions. The HRA department takes care of all these issues for the employees. The job is not high-tech or very glamorous but is company critical, which is why the position reports directly to the President and CEO.

Figure 3.31 is an illustration of the human resources and administration function.

ECONOMIC DOWNTURN AND THE ROLE OF HRA

The HRA department's role is to ensure that the employees' well-being is preserved. It acts as advocate for the carrier and for the employees. Both parties' needs must be met in a professional and compassionate fashion. It is an empty company whose benefits and overall treatment of employees cannot be used as a selling point. Unfortunately, more often than not, improving employee benefits is done only when the company is losing staff and is trying keep existing people while attracting new staff.

The recent downturn in the telecom market has left companies in a "position of power." In other words, it's an employer's job market. Companies may be downsizing, but they are hiring people to fill key spots. It goes without saying, that as companies downsize, the company doing the downsizing has a

Employees make the company

- Benefits
- Administrative Support

FIGURE 3.31 Vice President of Human Resources and Administration.

moral (not legal) obligation to assist employees in preparing for a change in employment status. The HRA department now administers programs generically called "outplacement programs," and does so for a number of reasons. First, it is the right thing to do and second, it may forestall an outraged employee's suing the carrier for unemployment compensation. Telecommunications carriers have been running employee outplacement programs since the mid-1980s.

The purpose of an employee outplacement program is to prepare the employee for life after Carrier A (for example) and for the task of finding a new job. Over the years, outplacement programs have come to perform the following types of tasks for the displaced employee:

- Teaching the displaced employee how to write a resume
- Demonstrating how to write a cover letter for a resume
- Advising on the content of the resume—many people do not know how to write a resume or even how to state past succinctly accomplishments
- Providing resume typing and photocopying services
- Providing the equipment (desktop computers and printers) to type resumes—most displaced employees will do their own typing and many do not have their own home computers
- Showing the displaced employee how to
  - Network with friends, family, and colleagues to seek a job or job leads
  - Respond in a job interview
  - Dress for a job interview
  - Act during a job interview
- Providing critiquing services for post-job interview reviews—many candidates will go to the outplacement center in order to discuss their job interviews with counselors
- Some carriers have even gone as far as providing job leads
- Providing access to financial advisors to assist the displaced employees during what could be a long period of unemployment

Outplacement centers are usually managed by external firms contracted to the telecom carrier. These contracts enable a displaced employee to use the center for a specific time period. The employee is allowed to go to the center as often as he wants. This is an important point to bear in mind. Losing one's job is frightening. The period of unemployment will cause the displaced employee and his family enormous stress. The outplacement center becomes an unofficial emotional counseling center. Displaced employees can sit and console themselves (and others) by being with colleagues who have been released by their common employer. In some cases, a displaced employee will find a job through a candidate who has found a job. Figure 3.32 is a depiction of the outplacement center's functions.

EMPLOYEE BENEFITS—PROMOTING THE COMPANY

Paying higher salaries is not the only way to attract employees. Carriers are extremely competitive and have used employee benefits as a way of attracting new hires. These benefits may include medical, dental, and vision plans as well as paid leaves of absence, and short-term and long-term disability plans.

Outplacement for displaced employees

Career development

Education

People networking

- Operations
- Network management
- Operator services
- Software development

FIGURE 3.32 Human resources and outplacement.

The point is that employee benefits are a way of convincing people to come and work for a company. Many carriers tend to pay the same annual salaries. The differentiator for prospective new hires is the employee benefits. If you have a family, medical, life insurance, dental, and vision plans are extremely important. Medical insurance is so costly that a well-paying position probably will not be able to cover the cost of medical care; medical benefits are often a major deciding factor for job seekers. During economic downturns, people may be desperate, but not so desperate that they do not want or need good benefits.

Pensions are almost things of the past with most telecommunications carriers. Today, most carriers, have employee 401(k) plans with company contributions, as well as company savings plans and company stock plans. A telecommunications carrier that still has an active pension plan may be very attractive to prospective employees.

FISCAL ACCOUNTABILITY

All senior executives and managers are held to a standard of fiscal accountability. Accountability is what sets apart those managers who just make budget decisions from those who are held accountable to investors and a board of directors. Every member of a competent team of senior managers has some level of understanding of every other member's job function. If there were no interdisciplinary knowledge, senior management could never function as a team.

Fiscal accountability is the common bond between all members of senior management. Not all senior management interfaces regularly with other senior managers. However, all interface regularly while managing budgets and meeting the organization's financial goals. Senior management staff meetings tend to focus on status, major decisions, task assignments, and company finances, since the financial health of the company is a concern for all senior managers.

SUMMARY

Carrier organizations vary slightly from company to company, although the variations may be small. All telecom carriers perform the same function: providing telecommunications services. Since all carriers are organized along functional lines (functional departmentation), all carrier organizations look similar.

The goal of a telecom carrier is to provide service to its customer base. The company needs to be managed in a way that produces an acceptable return on investment. However, the company must be organized correctly. One cannot simply combine departments to cut head count. Each function requires different skill sets. Just because someone is an engineer does not mean that he actually knows how to handle a tool to fix a cell site or even an air conditioner. There are specialists in medical benefits plans and dental plans. There are people who know how to manage the network, although they know nothing about why the network was designed as it was. Network operations people are not traffic engineers.

Theoretically, a company is organized to optimize the use of its human resources pool. Optimization means running the company with the right person for the right job. The next chapter will look at telecommunications vendors' organizational structure.

CHAPTER FOUR

THE TELECOMMUNICATIONS VENDOR—HARDWARE, SOFTWARE, AND SERVICES

The telecom vendor can make hardware, write software code, or sell services to a carrier (or another vendor). (Note that I often interchange the words vendor and manufacturer when discussing telecommunications companies that make hardware, software, or sell services to a carrier or another manufacturer.) The telecom hardware manufacturers build and resell a variety of equipment. Some of the equipment can only be used to provide telecommunications services, while other pieces of equipment have high reusability in other industries. The following is a list of the types of equipment used in the telecommunications environment:

- Trucks
- Switches—ATM, SS7, etc.
- Cabling—for internal wiring
- Transmission facilities
- Databases

- Frames
- Desktop computers
- Laptop computers
- Routers
- Cable racks
- Hand tools
- High impact glass (for network operations centers)
- Fire suppression systems
- Security systems
- Cell towers
- Microwave antennas
- Generators
- Batteries
- Mobile handsets

Figure 4.1 is an illustration of the kinds of hardware sold in the telecommunications space.

FIGURE 4.1 Telecommunications hardware.

As noted in Chapter 1, software is like hardware. There are different kinds of software. Software can be defined as a set of instructions given to a computer or some piece of hardware. It can be grouped into two basic categories: source code and object code.

In the telecommunications space, makers of software sell to terminal device manufacturers, users, and service providers. A terminal device includes any device with which a user directly interacts. Examples of terminal devices are mobile handsets, desktop computers, laptop computers, handheld personal data assistants, and pagers. A user typically purchases a terminal device. Recall that the service provider also purchases software to operate its switching or routing equipment and its "back office" systems such as customer billing, network element management, and network monitoring.

Providers of software to the telecom segment can include companies typically associated with telecommunications or even those not so associated. These companies may be providers of business software for word processing. Other companies sell software that developers may use to create applications. Software tools originally intended for the Internet Web page business space has found its way into the hands of telecommunications vendors and carriers. Figure 4.2 is an illustration of the kinds of software sold in the telecommunications business space.

As I noted in Chapter 1, another sector of the telecommunication vendor business is the vendors who sell services to one another and to carriers. The hardware and software sectors of the telecom business often go outside their companies in order to obtain a variety of development resources they normally do not keep on staff. The resource gap is filled by consultants and development companies (sometimes generically known as development houses). The services rendered can include:

- Hardware design
- Software development
- Human resources—many companies now hire external firms to manage their human resource departmental needs; this practice is called *outsourcing*

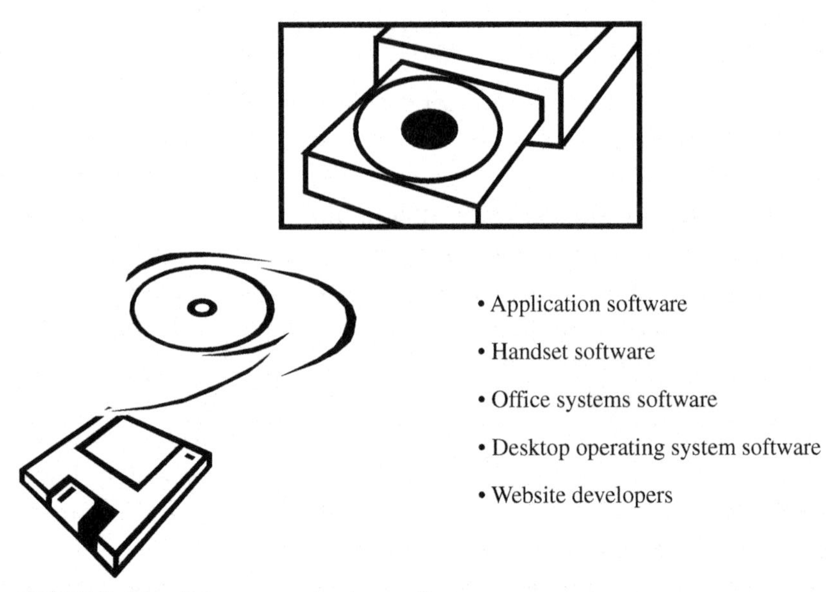

FIGURE 4.2 Telecommunications software.

- Marketing consultation
- Hardware and software testing
- System integration (system assembly)
- Warehousing
- Delivery
- Technology consultation—if the company does not possess the expertise, it will either hire someone who does or hire a trainer to teach the technology to the existing staff.
- Operations outsourcing

A company that creates hardware or software will often outsource as much as possible and focus on what it considers its core specialty. There is always a financial imperative behind the decision on whether or not to outsource. Sometimes the companies that render a service to the manufacturers also provide services to the telecommunications carriers. The concept of outsourcing was popular in the early 1990s and, given the recent (circa 2000) economic downturn, has become popular again.

OUTSOURCING

One can outsource aspects of finance, operations, directory assistance, and even customer care. In *Telecommunications Internetworking* I delve deeply into the concept of outsourcing. Telecom carriers have been outsourcing for many years. For the purposes of this book, we should simply note that the drive to reduce operating costs has caused many carriers to go outside their companies to have parts of their business run less expensively by other companies.

Outsourcing requires that the company performing the function have an intimate understanding of the nature of the activity it is performing on behalf of the carrier. Let us call the company performing the outsourced function Outsourced XYZ. Outsourced XYZ performs the function of directory assistance for a carrier. Some of the questions that Outsourced XYZ needs to have answered are:

- Does the carrier want regional accents used?
- How does the carrier want each call answered? In other words is there a "banner opening line," like "Good day, this is Carrier ABC"?
- How does the carrier want each call ended? Is there a "banner closing line," like, "Thank you for using Carrier ABC, and have a great day"?
- Does the carrier want Outsourced XYZ operators familiar with the local weather?
- How long does the carrier want each call to last with each customer?
- How often does the carrier want a performance report?

MULTITUDE OF VENDOR TYPES—IMPACT ON ORGANIZATIONAL STRUCTURE

As the reader can see, there are a multitude of vendor types, performing many kinds of functions for customers. This multiplicity has an impact on the kinds of organizational structures that need to be in place in order to run the business and meet the customers' needs.

If we went from vendor to vendor, we would find many superficial differences. Groups of engineers that perform some kind of radio engineering function would have different names in different manufacturing shops. For example, there are two big manufacturers of cellular switches and base station equipment. In Vendor A, one department is responsible for designing the switch and all enhancements. The same department is designing a new kind of multifunctional switch (called a softswitch). The department has divided its organization into two groups: one is responsible for the old switch and the other is responsible for the new switch. In Vendor B, one department is responsible for designing an existing switch and all its enhancements. However, Vendor B has created another department (with its own vice president), which is responsible for all new switching technology.

A software vendor makes operating systems, word processing, and spreadsheet software. Another software vendor makes word processing, spreadsheet, and drawing program software. How are the two vendors organized? Is the software coding group responsible for the spreadsheet and word processing coding? Do the software packages interact with one another? Is it possible to port data from the word processing program to the spreadsheet or drawing programs?

Are hardware manufacturers organized the same way software companies are organized? The answer is yes and no.

Vendors organize their companies along product lines, not functions. As I had noted in Chapter 3, product departmentation is the norm for manufacturers, both hardware and software. However, we find upon burrowing deeper into any given department that functional departmentation occurs so that human resources can be used across multiple activities.

BASIC ORGANIZATIONAL FUNCTIONS

From the 100,000-foot altitude perspective, all telecommunications carriers and vendors are organized in the same way. The types of skill sets or functions needed to run a telecommunications company are the same across any company type. The skill areas are:

- Operational/maintenance
- Technical
- Administrative
- Management
- Business—sales and marketing
- Finance
- Human resources

As I noted in Chapter 3, the categories are very broad and applicable to any company, and the functions within each category are linked to providing a service. The same can be said for a company manufacturing hardware or software. The service is the product. To some extent, the more one generalizes, the more applicable the organizational structure can be. Remember, this must be approached from a high-level perspective. Imagine a house under construction. First, you put in either the slab or the foundation. One foundation or slab looks like any other slab or foundation. Next, you frame out the house. The house framework is the skeletal structure upon which everything is built. What defines the style and appearance of a house are the decorating details.

Figure 4.3 is an illustration of the basic organizational functions of a telecommunications vendor, as seen from a high-level viewpoint.

ORGANIZING THE VENDOR

It is necessary to decide how to divide the labor or decision processes so that a company can operate effectively. I described the concept of departmentation in Chapter 3. Product, function, territory, time, quality of service, customer segment,

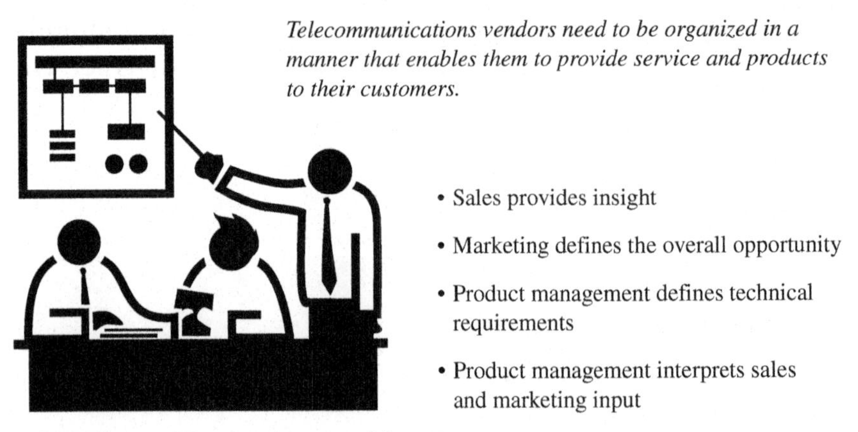

FIGURE 4.3 Basic organizational functions.

equipment type, and alphanumeric organization are all parameters of an organizational principle called departmentation.

Telecommunications vendors need to be organized in a manner that enables them to provide service and products to their customers. Vendors always work toward meeting stated and perceived carrier (and sometimes other vendor) expectations. The biggest challenge for the manufacturer is understanding what the customer has said and what the customer has not said. Figure 4.4 is an illustration of departmentation

FIGURE 4.4 Departmentation and telecommunications.

DEPARTMENTATION

For the purposes of this chapter, I the concept of departmentation is a principle in which a company creates organizations based on certain parameters or key criteria. As I noted earlier, the forms of departmentation are:

- *Product.* Large manufacturing companies are typically organized along product lines. An example is the automobile manufacturer. Within the product line the group may be organized along functions. Product-organized companies like automobile manufacturers are able to create self-contained product organizations. For example, a specific model of car is assembled in a specific plant. All functions required to assemble the car are in one factory. The automobile manufacturers have used product departmentation effectively for decades.
- *Functional.* Functional departmentation refers to activities such as engineering, marketing, and finance.
- *Territory.* Territorial departmentation refers to regions or divisions of a multinational company that operate as separate companies but still report to a single headquarters.
- *Time.* Time departmentation refers to shift work.
- *Quality of Service.* Quality of Service departmentation refers to classes of service such as first class and coach on a plane or train.
- *Customer segmentation.* Customer segmentation focuses on specific customer segments. In a telecommunications company, the customers may be broken down into residential or business. A bank may have a loan department focused on farmers and another department focused on small businesses.
- *Equipment type.* Automobile manufacturers may have all drill press work done in a single area while welding is done in another area.
- *Alphanumeric organization.* Alphanumeric departmentation is common in many companies. In a telecom carrier, customers

with last names beginning with the letters A through L are apportioned in one direction, while everyone else goes the other way.

Departmentation parameters can be used simultaneously to varying degrees. However, the company is usually organized around one particular parameter. The most popular forms of departmentation in business, and especially the telecommunications industry, are product and functional departmentation. Telecommunications carriers are functionally organized. However, vendors are organized along product lines; the organizational focus is product departmentation.

PRODUCT DEPARTMENTATION

Large manufacturing companies are typically organized along product lines. An example is the aircraft manufacturer, organized along models of airplanes. In other words, one organization is responsible for designing, building, and maintaining a jumbo jet. Another organization in the aircraft company is responsible for building a smaller jet used for shuttle service between cities that are close to one another. Deeper within the organization, the company will functionally organize the various departments. What we have in this situation is the aircraft company executing product departmentation at the primary level and all other levels using both functional and product departmentation. Figure 4.5 is an illustration of product departmentation.

Telecommunications vendors are also product oriented. Remember, products are designed and built to meet specific customer needs: no single telecommunications product does everything, because the cost of that switch or product would be too high for any customer (carrier or vendor). Telecommunications vendors will have all hardware work associated with a particular switch or router handled by one facility. The software efforts may be handled in another location. Further, the software groups are probably dedicated to the specific product line.

FIGURE 4.5 Product departmentation.

At a primary level, a telecommunications vendor may choose to organize along product lines. The product form of departmentation is adaptable to very large, complex organizations. Telecommunications technology is complicated. Product departmentation enables vendors to break down technology initiatives into self-contained, manageable, smaller product organizations. Functional departmentation is used in cases in which skill sets and facilities can be reused with little retraining. In effect, the advantages of both large and small size can occur simultaneously in one large organization. Figure 4.6 is an illustration of product departmentation and functional departmentation.

Specialization is an attribute of product departmentation. Organizations managed along product lines are comprised of specialists. These people are often dedicated to the product. Specialties may include areas such as batteries, welding, chip design, chip manufacturing, product cleansing, electrical grounding, and connectors. Each organization exists on its own and maintains its own profit-and-loss statements.

Examples of Product Departmentation:	Examples Functional Departmentaion:
• Handsets	• Assembly
• Terminals	• Switch Design
• Switches	• Maintenance
• Routers	• Legal
	• Public Relations

FIGURE 4.6 Product departmentation and functional departmentation.

What we see is that product departmentation enables the company management to exert greater and finer control over the company than if functional departmentation were used. Since the organizations are self-contained, they are adaptable to accounting-control techniques and management assessment. Further, product organizations can be easily added or dropped, with a minimum of impact on the rest of the company. To some extent, the manner in which telecommunications vendors work has enabled them to adapt easily to the economic downturn by reducing expenses via product-line elimination. Figure 4.7 illustrates product departmentation's adaptability and friendliness.

PRODUCT DEPARTMENTATION: DIVERSIFICATION AND MARKET RESPONSE

Many telecommunications vendors provide at least two different products or services. A vendor may specialize in optical

Companies organized around the principles of
Product Departmentation are:

- Human resource heavy
- Large numbers of specialists
- Resources can easily added or dropped

FIGURE 4.7 Product departmentation—adaptability.

equipment but provide at least two different versions of a piece of equipment. Diversification enables the manufacturer to meet a variety of needs stated by the customer as well as those projected by the vendor's sales and marketing teams.

The ability to add and drop product lines enables a vendor to quickly shift its focus, financial, and human resources into new areas of business. This is a major advantage of the product departmentation: for example, a telecommunications vendor may manufacture and sell time division multiplexed switches, routers, databases, customer premise equipment, cellular base stations, and transmission facilities. One day, the vendor realizes there is no longer any growing business in time division multiplexed switches and decides to enter the softswitch market. The vendor can do one of two things: nurture the necessary skill sets to design and build a softswitch or buy a company already in the softswitch business. Given the short market time windows and the long lead times needed to nurture in-house talent, it is simpler to buy a company already in the softswitch business. Dumping the existing product line presents a problem only for the customer who wishes to have continued support. The vendor can maintain a group for maintenance of the old product, but the bulk of the people and maybe even the facilities (buildings, test equipment, manufacturing equipment) can be easily removed from the corporate ledgers with minimal impact on the operation of the other organizations.

The vendor buys a company in the softswitch business and integrates it into the internal communications process. If that company has its own manufacturing facilities, the purchasing vendor can either integrate the facility or get rid of the manufacturing plant. What is important to the purchaser is that it now owns the intellectual property and employs the people who created the intellectual property. Furthermore, an existing softswitch company also has a financial track record that enables the purchaser to gauge the impact of the new product on the purchaser. Corporate culture integration and senior management egos aside, integrating an existing company is far more operationally and financially attractive than spending time building a new organization using nothing but existing internal resources. An added bonus would be the target company's financial track record. If the target company had positive cash flow, the acquiring company would add that cash flow to its revenue stream. These are the reasons why mergers are so attractive.

Having a diverse product line increases a vendor's chances of survival in a rapidly changing marketplace. Going the mergers and acquisitions route is one way of quickly achieving diversity. Another way is planning for it from the beginning and using in-house talent to nurture the additional opportunity.

Regardless of how one wishes to achieve diversity, if it were not for the fact vendors are organized along product lines it would be a very difficult task to financially or operational integrate a new product line. Companies organized under the rules of product departmentation can be thought of as being comprised of a collection of entities that can exist independently of one another but gain strength by working together. The attributes of such a collective are:

- Branding power—market power in using a well-known company name
- Buying power—the larger the company, the greater the buying power
- Incremental growth provided to the corporate cash coffers by each part of the company
- Economies of scale in using common tools and resources

Think of product departmentation as a completed puzzle within a completed puzzle. Each puzzle can stand alone, but when they are put together, the total can result in a larger but more detailed picture. Figure 4.8 is an illustration of product departmentation.

VENDOR ORGANIZATIONS

Despite the fact that product and functional departmentation are two completely different organizational forms, they follow

Product Departmenation enables a company to structure itself as it were many companies with a single company.

FIGURE 4.8 Product departmentation—puzzle within a puzzle.

the same template of job functions. In other words, at a primary level, product departmentation seeks to organize along product lines while functional departmentation seeks to organize along functional lines. However, at the secondary level (one level lower), both product departmentation and functional departmentation seek to have the same primary functions performed:

- Marketing
- Sales
- Customer care
- Engineering—design
- Operations—the factory worked for a vendor and network managers for a carrier. These are the people on the ground level.
- Finance
- Legislative and regulatory
- Legal

Figure 4.9 is an illustration of these primary functions.

As I mentioned in Chapter 3, we should note that these primary functions usually end up being organized in senior executive positions according to the following vice presidential categories:

- Marketing
- Finance
- Operations
- Sales
- Technology

Carriers and Vendors operate on the same principles.

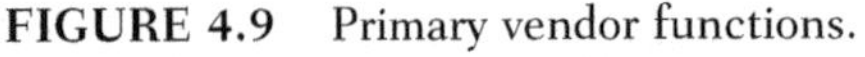

FIGURE 4.9 Primary vendor functions.

- General Counsel and Regulatory
- Vice President of Public Relations
- Chief Technical Officer
- Chief Financial Officer
- Vice President of Sales and Marketing
- Chief Operating Officer
- Vice President of Human Resources and Administration

All these positions report directly to the President and CEO. Figure 4.10 is an illustration of the senior management

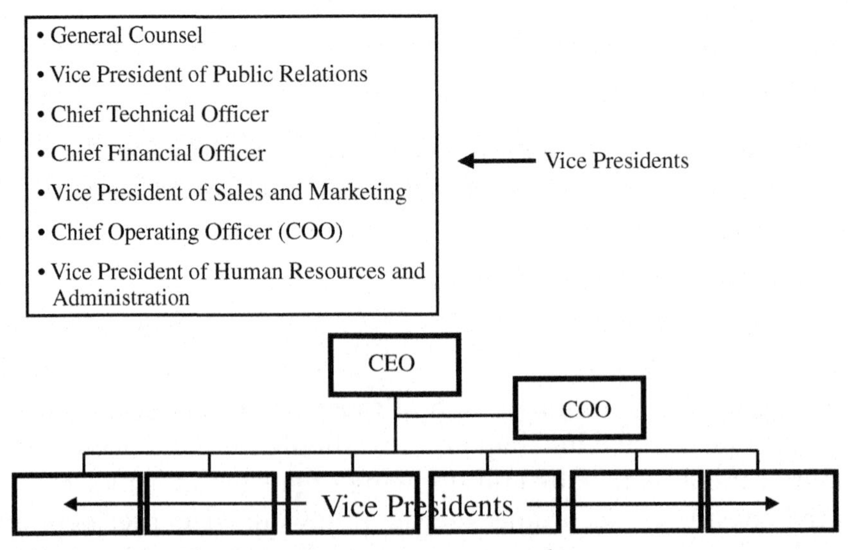

FIGURE 4.10 The Senior Executive organization chart.

structure. The following section will delve deeper into this structure. To understand how and why functions are organized as they are, we need to first understand how and why the senior decision makers are organized as they are.

The vendor organizational positions are not different from the corresponding carrier positions, except that the product the company produces is different. The functions are the same because both the carrier and the vendor are performing a service or building a product and selling it. Both company types are for-profit businesses serving customers. Both are seeking to make money. The mechanisms needed to operate a profitable business are the same.

As one burrows deeper into the company's organization one approaches the core of the company. At this point, the differences between the companies are more noticeable. A vendor that makes switches is different from a telecommunications carrier. A vendor that makes routers is different from a software vendor. For the purposes of this book, we are going to make the analysis at a high level.

SENIOR MANAGEMENT

Senior management is a generic term used to refer to the most senior-level decision makers of the company. Typically, very large companies divide the senior management teams into executive and senior executive teams, because in a large company, the sheer number of employees requires the creation of more than one layer of executives in order to divide the management responsibilities among more-senior managers.

The senior management team in a vendor is divided into the following areas of responsibility:

- Legal
- Regulatory
- Legislative
- Finance
- Technology

- Operations
- Sales and marketing
- Customer care
- Human resources
- Public relations

When we look at all the functions that reside in a vendor, ranging from installation to billing, each function can be generalized enough to find a place in one of the categories in this list.

These categories are used as a basis for the creation of organizational structures, which take the form of the following departments or organizations. The vice presidential positions created are:

- General Counsel and Vice President of Legislative/Regulatory—legislative, regulatory, and legal
- Vice President of Public Relations
- Chief Technical Officer (CTO)—technology and network operations
- Chief Financial Officer (CFO)—all corporate financial matters and customer invoicing
- Vice President of Sales & Marketing
- Chief Operating Officer (COO)—customer care
- Vice President of Human Resources and Administration

These functions are described in detail in Chapter 3 and are generally the same for a vendor. Figure 4.11 is an illustration of the vice presidential positions that operate in a vendor.

PRODUCT LINES

Even in companies that are product departmentation–based, functional organization is essential. Among small vendors, a company may be a startup and therefore have only limited

- Finance
- Technology
- Marketing
- Sales
- Operations

Vice presidential organizations in the vendor community are focused on producing product that meets the customers' needs and selling product.

FIGURE 4.11 Vice Presidential organizations—vendor.

resources. In small companies, the vendor may be oriented by function rather than by product. However, even small companies need to organize along product lines at some point, as a matter of resource efficiency and focus. Developers of hardware and software need to focus on the individual product. The various product lines may be based on a common underlying technology, but the individual product still needs to be given particular attention.

One of the important aspects of any vendor is the general flow of information between the customer segment (or individual customer) and the vendor. The process of a vendor's product creation is fascinating and it represents the essence of the vendor. The remainder of this chapter will focus on the internal processes of product creation.

THE PROCESS OF CREATION

The vendor needs to approach the development of product (whether it's a service or a physical product) with a clear focus on meeting customer needs. The biggest challenge facing any telecom vendor is making sure it does not fall in love with its own product and begin to believe in its own marketing. This is called "drinking your own bath water"; you immerse others in the product marketing story (the water) and then you do the same thing and believe in the hype. This is taking faith in products to dangerous levels because a vendor can lose sight of what is important: selling product. The vendor must be able to walk away from one product to create and sell another.

The process of creation is an orderly and regimented one. It begins with the marketing department, which first needs to identify a market opportunity. The marketing department then examines the marketplace to determine if the proposed product matches projected customer needs. The sales department performs a sanity check of the marketing department's efforts.

This is followed by a well-thought-out product definition proposal. The sales and technology teams have simultaneous input into the product definition. The sales teams have direct access to customers. The input gained from the customers will enable the technology people to design the product to meet the common needs of multiple numbers of customers. The formalized process is comprised of defined and yet fluid processes designed to check constantly on each other. It starts with customer input and market research. The sales and marketing organizations are key to ensuring the company makes the correct product. Vendors are rarely given a chance to make a mistake in the marketplace. Figure 4.12 is a representation of the creative process.

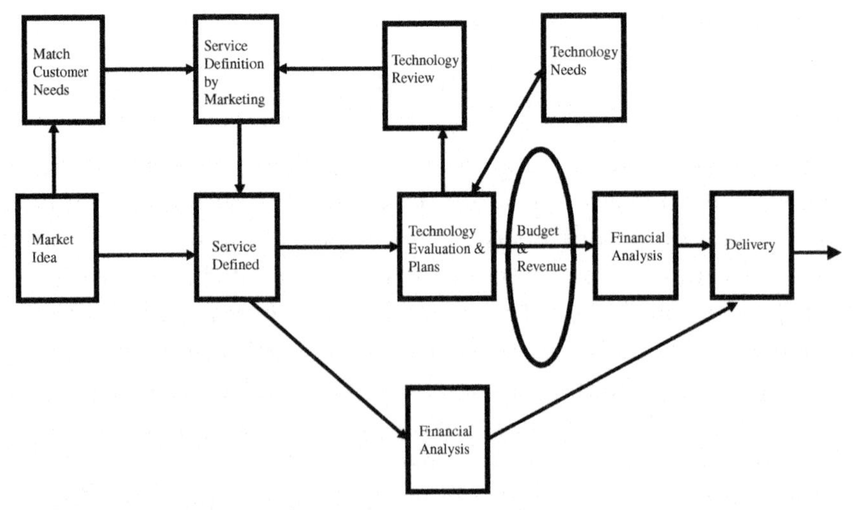

FIGURE 4.12 Process of creation.

THE CREATIVE PROCESS—ORGANIZED CREATION AND CREATIVE MANAGEMENT

Despite the belief that technology creation is or should be characterized as unfettered and unbridled creative technical energy, successful commercial vendors do not allow this to occur. Unless it is the old Bell system, when competition was nearly nonexistent and money was plentiful, the vendor is not going to allow any type of unmanaged research or design work. Manufacturers and vendors call this attitude different things, but each one establishes a rigid process of measuring and benchmarking the product development process. We will call this the d*evelopment loop*.

The loop starts with marketing and sales, who work together. The marketing department is working to define the marketplace needs. It will hold focus group meetings to get market input. Whether the vendor is a startup or an established company, the next step begins before the research and market definition work is completed. This step involves interaction with the company's technology and sales organizations. The marketing department needs input from sales and technology to determine if the marketing work is on target or at least on the right track. The sales people have a real-world sense for what their customers are seeking. The technology people know what can and cannot be built within the various financial and time constraints established by the investors, board of directors, and stockholders. Figure 4.13 is an illustration of this part of the development loop.

Note that there is a feedback loop among the sales, technology and marketing organizations. As the product is being

- The vendor makes a decision as to what it will sell. Once this decision is made, a process of continually checking back with the customer or target customers needs to be established.
- Customer input and information feedback.

FIGURE 4.13 Development loop—initial steps.

developed, the customer-contact groups in sales and marketing are feeding back information to the technology development people in order to ensure the product is fine tuned to meet the customers' most current expectations. Remember, the early stages of product creation are critical. Once the vendor embarks on a particular product path, the financial and human resources needed to develop the product will be locked in and difficult to change quickly. There is absolutely no room for experimentation with an idea. A company cannot expect to stay in business for very long if it changes the entire product in the middle of the development process. The vendor should expect to make adjustments to the product or the product mix, however to expect a company to have the financial resources to make major changes in the nature of the product or product mix is not realistic.

Successful vendor startups are particularly good at this part of the creative process. A startup has limited financial resources and therefore has to make a series of good decisions or, as some may say, decisions that are not totally wrong.

THE CREATIVE PROCESS—PRODUCT MANAGEMENT

At some point, the vendor makes a decision about what it will sell. Once this decision is made, a process of continually checking back with the customer or target customers needs to be established. This continual process of checking and feeding customer and market information into vendor's various technology and business components is critical. Often the marketplace or the customer will change its mind about what it wants, and the vendor needs to be able to respond, with minimal impact on the product development process.

Product management is a group within the marketing organization. Its job is to act as facilitator among the sales, marketing, and technology departments. The product management group is the technical interpreter of what the sales department hears from the customer, what the marketing department believes it is hearing from the overall customer segment, and what the technology department believes it can build. The product management department is especially important in translating customer needs into technical requirements that the technology department can use to design the product. Figure 4.14 is an illustration of the product manager's job, which is to manage the overall development process of the product. Keep in mind that no product is developed without ongoing input from every department in a vendor organization.

EXAMPLE OF PRODUCT MANAGEMENT PROCESS

The product manager is responsible for managing the entire creation process for a particular product. Let us say that the vendor is entering the softswitch business. The CEO and executive team decide that the company should shift its resources from a time division multiplexed switch, but the decision needs to be studied before it can be finalized. The study looks into the details of the CEO's desire to determine the feasibility of moving ahead with this wish. The Vice President of Marketing assigns a product manager. The product manager sets up a first meeting with all the departments: sales, product marketing,

- The product manager's job is to manage the overall development process of the product.
- Product management is a step-by-step process.
- Product management requires a top-down and bottoms-up approach.

FIGURE 4.14 Product management.

technology, finance, manufacturing, business development, market management, and installation. At this first meeting, the product manager obtains commitments from the various departments to perform the feasibility study. The product manager establishes a timeline for completion of the study.

For the purposes of this example, we will say the study comes back with the following recommendations and comments:

- The company can develop its own softswitch within time frame XYZ.
- The company will need to hire N number of people in the following areas:
  - Software
  - Hardware
  - Product Marketing—creating the industry advertising pitch
- Purchasing software development tools will be required.
- The project will cost $xxx.
- The project will reach the field testing stage in one year.
- The softswitch product will be ready for commercial availability six months after the field test.
- The sales department has identified 10 potential customers. These have made no commitment, but have expressed strong interest in the product concept.
- The market management group has determined that the company may able to expand its market into another country within days of the product's commercial availability.

Upon reading the report's recommendations and comments, the CEO and the executive team decide to move forward on the softswitch product. The product manager then sets up a second meeting. The project has now reached the first part of the execution phase. At this second meeting, the product manager meets with the team of people who will be dedicated to ensuring the creation of the company's softswitch. The team is comprised of departments that were a part of the first meeting: sales, product marketing, technology, finance, manufacturing, business development, market management, and installation. Overseeing the entire effort is a single management sponsor, usually a senior-level person within the marketing department.

The job of the product manager is to ensure that the customer's needs are met by the product. The salesperson will seek live and continuing input from the target customers. The product marketing person will gather market trending data to enable the product team to project future capabilities of the product. A projection of future needs will enable the technology department to put the necessary placeholders in the software for easy future enhancements. The market manager seeks to build overall market excitement based on the work of the group. This manager is not a day-to-day participant, but rather someone who sits and listens. The finance person is there to ensure that the company's financial objectives (product costs and margins) and the project's financial parameters (budget) are met. The company should already have a standard warranty and repair policy in place that is applicable across all product lines. However, the softswitch product may need to be treated differently and a different warranty and repair policy will be needed. The business development person is there to help find ways of leveraging corporate business relationships or creating new business relationships that enable the company to optimize the way in which the product is being developed and the way in which it can be sold. The installation people are there to ensure that the product can be installed and maintained with minimal cost to the company. In other words, the installation force will be held accountable for field maintenance of the product after the product is sold and installed. The company does not want to spend a large amount of time or money performing this function. Even more important, the product has to fit the customer's floor space requirements. This point in the product management process is often called the *development phase*. Figure 4.15 is an illustration of the product management process.

One of the goals of product management is coordinating the actions of all departments to ensure timely delivery of the product.

The Completion Phase is the point at which the product has gone commercial.

The Commercial Stage is when the product is ready to be sold to customers.

FIGURE 4.15 Product management process.

Figure 4.15 includes a phase known as the *completion phase*. This is the point at which the product has gone commercial; it is ready to be sold to customers—this is the handoff to sales. Technology that has gone through the entire development process and passed through the completion phase is considered a product. Technology being sold to customers for commercial use is considered a product and not technology.

The product management process has places in which feedback on certain issues, such as finance, are required. There are also provisions for software checks. Note that once a decision been made and the project moves on to the next stage, the company does not retreat or look back. If the product management process is executed properly, there should be no need to double-check results. As the product is being tested, technology assumptions are being checked. If the sales and marketing people are performing their jobs properly, customer- and market-related input is always sought after each stage of the process. Of course, if there is a radical (and unanticipated) market shift, product changes will be necessary even if the product has reached the completion phase. When unanticipated changes occur, the sales and marketing people will find a way of living with what they have been given. However, the required changes will probably make their way into the product's next release, in order to avoid delaying the introduction of the product's first release.

PRODUCT MANAGEMENT: MEDIATION, FACILITATION, COORDINATION—HITTING THE MARK

The most difficult task in a vendor shop is coordination between departments. In reality, we are dealing with a company's specific agenda; but departments also have their own agendas. One of the goals of product management is coordinating the actions of all departments to ensure timely delivery of the product. Missing a deadline or misunderstanding the market's needs can be disastrous to a vendor. To be the first in the telecom industry to sell the first telecommunications widget means a great deal from two perspectives:

- The vendor is the first to set the price. Rather than a competitor's setting the price to meet its financial needs, you can set the price to meet your needs.
- The public relations right to say your company is the first.

"Missing the mark" on a product usually results in lots of blame being handed out to lots of people. Finger pointing in the midst of such a disaster is common. Unfortunately, product creation is a process that can be characterized by tension and frayed nerves. The need to coordinate closely with others is paramount to the survival of a vendor, especially a startup company. The job of product management to coordinate and facilitate discussions between departments is paramount. In some vendor shops, the position of product management is so critical that it is given a separate vice president. In very large manufacturing or software companies, the position of product management is accorded the status of officer.

There can be no room for empire building. Unfortunately, empire building occurs far too often in startups. When empire building occurs during the creation of the company's very first product, the CEO must stop it immediately.

Small companies have no scale or market dominance to use to cover up inefficiencies. Mistakes cannot be hidden. Coordination is a critical and often understated element of success for a vendor. Given the state of the telecommunications market in 2000–2002, meeting market needs is important. The old sales adage of "sell the customer what you have and not what they want" no longer applies. The telecom carrier community is aware of every trick a telecommunications salesperson has ever tried. Selling to a carrier's competitor is a tool that usually works. In this scenario, say that a carrier is not sure if it should proceed with a particular service using a particular product technology. The carrier cannot justify the need for the product, therefore the vendor cannot make the sale to this carrier. However, the vendor has found another carrier that wants to try out this new technology. Therefore, the vendor sells to this other carrier. The first carrier the vendor approached (who could not decide) sees what has occurred and

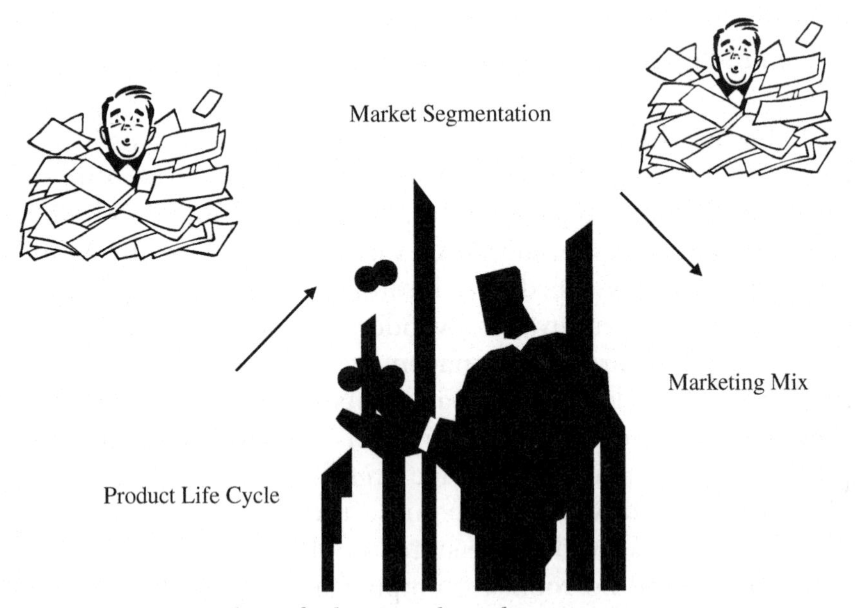

FIGURE 4.16 Mediator, facilitator, and coordinator.

now wants the same product; this is the instant business case and the instant sale. Figure 4.16 is an illustration of product management's role as mediator, facilitator, and coordinator.

In the product management process, the danger exists in focusing too much on the process for the sake of the process, and forgetting that the goals are selling product, making money, and satisfying customer needs. Positioning a product in the marketplace is more art than science. It entails communicating a message and image to the vendor's customer base. The matter of market positioning is discussed in Chapter 5.

PRODUCT DEFINITION

Defining the service is always a major challenge for a vendor. The technology development process runs in parallel with the service definition process and therefore the whole effort is a major coordination job.

The basic approach to defining a product is a three-step process, focused on defining not the actual service but rather aspects of marketing. Remember, marketing is about defining the product and the environment in which it is sold. The three steps are: market segmentation, marketing mix, and product life cycle. Figure 4.17 is an illustration of the product definition process.

Market segmentation is the process of analyzing the markets in terms of discrete parts or segments. It assumes that different parts of the marketplace have different needs and requirements. The challenge for the vendor is to identify the market segments properly. These segments can be broken down into any number of categories. For example:

- Market segmentation – defining the market
- Marketing mix – variables controlled by the company
- Product life cycle – useful and profitable life span of the product

FIGURE 4.17 Product definition process.

- Small, medium-sized, or large carrier
- Business customers of the carrier
  - Small
  - Medium
  - Large
- Rural carrier
- Urban carrier
- Residential customers of the carrier
- Carrier that owns fiber
- Carrier that owns transmission facilities
- Carrier with its own highly skilled technology department vs. one that is not highly knowledgeable
- Carrier that has a small maintenance force—this could be an opportunity for the vendor's own installation forces. Installation is not free.
- Carriers that are wholesalers
- Carriers that are retailers
- Carriers that focus on the business segment only
- Carriers that focus on rural markets

Figure 4.18 is a depiction of market segmentation. The illustration highlights the multitude of ways in which a market can be segmented.

Examples:

- Rural carriers
- Urban carriers
- ISPs
- Wireline carriers in urban markets
- ISPs serving the residential community

FIGURE 4.18 Market segmentation.

These sample variables are called *demographics*. Demographics typically includes gender, age, income, occupation, family size, religion, gender of children if any, etc. Vendors take demographics in another direction and create similar profiles of their customers. I noted in *Telecommunications Internetworking* that the vendor can gather an enormous amount of information about the overall marketplace. How the information is used determines how the segments are identified.

The factors used to identify market segments are demographics, the carrier's geographic and business model, the serving area of the carrier, the history of the vendor with the carrier's management, service sold by the carrier, the carrier's spending habits, level of service usage, the size of the carrier's customer base, etc.

THE FOUR PS

The marketing mix is comprised of those variables and factors controlled by the carrier or vendor to plan, introduce, and sell the product. These factors are called the four Ps:

- Product
- Price
- Place
- Promotion

Product refers to the physical nature of the product and its perceived value. The product also includes the conduct and reputation of the vendor, and guarantees the vendor's offers. Product is very much a subjective factor, part perception and part fact. The people who build and sell the product are as much a part of the product as the material that went into constructing it. Many carrier managers (like any product purchaser) tend to favor products for a variety of personal reasons. These can cover a range of issues:

- Training of the CTO
- Experience of the CTO—professional experience, not experience with the product

- Experience with the product
- Carrier's past experience with the vendor's head of sales
- Carrier's past experience with any member of the vendor's executive management
- Product results from any carrier-run trials
- Color of the "box"
- Size of the "box"
- Flexibility of the product—whether it can be installed in a variety of environments or network configurations
- Ease of installation—many carriers say "no" to a vendor just based on difficulty of installation and high maintenance costs. Often carriers want to "plug it in" and "walk away."
- The behavior of the vendor's sales team—unprofessional behavior such as offering executives plane rides, nights on the town, expensive gifts, etc. is unacceptable. Once the sales team violates the unwritten code of conduct, carriers usually mark the company as trouble and one to avoid.
- The behavior of the vendor's field operations and installation team—groups considered unprofessional are viewed as possible sources of trouble in the future.
- The relationship between the vendor's and carrier's management teams
- The intrinsic value of the product. This takes all human perception and emotions out of the evaluation equation by answering the fundamental question: Why is the product valuable for the customer?

Price refers to the purchase price of the product, the terms and conditions of the payments, and discounts. Sometimes a carrier will decide on which vendor(s) to do business with based on product pricing. Sometimes the carrier will pay more because the value the product brings to the carrier is overwhelmingly greater than the value of other vendors' products. Pricing is a complex matter. On the one hand, there are vendors who seek to recoup their entire investment in a technology by making their first customer pay for it; other vendors recoup their investment in a technology by having multiple sales (customers) pay for it. The former situation usually results in no sale at all; you cannot make your first customer pay for all your development costs.

Carriers often issue Requests for Proposal (RFPs):

- To obtain data on how a network configuration should be designed
- To obtain data on how a service should be provisioned
- To obtain data on how to measure quality of services or how the technology is supposed to work in different environments
- To obtain sample product pricing data

More often than not, a vendor's pricing is requested in RFPs. Usually, a carrier will not perceive any differences significant enough to sway a purchasing decision one way or the other.

Place refers to how the vendor will sell and distribute the product. The vendor can sell directly to the subscriber base. It may even use the carrier as its agent to sell or lease equipment to another carrier. Vendors use a variety of methods, often innovative, to sell equipment. As I noted, they use their own carrier customers as their agents. In other instances, vendors will:

- Sell their equipment via larger vendors. The relationship is called an OEM (Original Equipment Manufacturer) relationship. The term is described in detail in Chapter 1.
- Lease their equipment only and not sell
- Finance their sale to a carrier in order to lock the carrier into a long-term relationship with the vendor. This can ultimately disrupt the financial health of the vendor if it is not done properly.
- Sell equipment via catalog
- Sell equipment via the Internet
- Make one big sale and then live off the upgrades and system enhancements

- Make one big sale and ask the carrier to refer the vendor
- Make one big sale and use that sale to intimidate the other potential customers into believing they must have the same product

Promotion refers to advertising and selling. To some extent, using a carrier sale to get another sale is an old sales tactic to promote a vendor's product. Nothing speaks louder than a sale a vendor can point to and say "see, your competitor bought and so should you." Telecommunications vendors who sell to carriers do not rely on advertising in trade journals, or even television, to get their clients. Rather, these vendors sell directly to carriers. The advertising in trade journals, television, or other mass media is for the sole purpose of creating "industry buzz," which affects the perceived value of the company. There have been instances where a vendor has decided not to advertise at all, resulting in carriers' believing the company had gone out of business. Because the carriers mistakenly believed the vendor had shut its doors, the vendor received no RFPs.

Vendors also participate in industry trade shows or conventions. In most industries a, big show booth usually leads the customer base to believe the vendor has lots of money and resources, and tremendous knowledge. Most key decision makers never make it to the convention or trade show booth floor, but staff always make it. The staff does the run through the convention floor, picking up brochures, visiting booths of vendors, having quick meetings on the convention floor, and attending seminars. The staff is affected by what it sees at the convention.

Vendors take every opportunity at industry trade shows to promote the company. Sometimes the promotion will take the form of dinners, dinner shows, golf outings, and bar hopping. In the end, promotion is about keeping the name of the company in the forefront of the industry and selling a positive image of the company.

Figure 4.19 is an illustration of the 4 Ps.

- Product refers to the physical nature of the product and the perceived value of the product.
- Price refers to the purchase price of the product, the terms and conditions of the payments, and discounts.
- Place refers to how the vendor will sell and distribute the product.
- Promotion refers to advertising and selling.

The Four Ps:
- Product
- Price
- Place
- Promotion

FIGURE 4.19 The four Ps.

PRODUCT LIFE CYCLE

Product life cycle refers to the life of the product. A product's life cycle is composed of four stages, which are all affected by the four Ps. The stages in the life of a product are:

- *Introduction.* The products initial entry into the marketplace. In the telecommunications business the introductory phase includes an intense period of selling, marketing, and educating.
- *Growth.* The period of time in which the product has caught on with the public, education is no longer a priority, and sales begins to grow.
- *Maturity.* Product maturity refers to the period of time in which the product's sales growth levels off due to market saturation.
- *Decline.* Product decline occurs when sales begin to decline. The decline is the result of either or both as a result of product obsolescence or product replacement.

CONTRACT SIGNING

A signed contract is what every vendor hopes for. It means revenue. It must be well written to protect the vendor and the carrier. There is more art than science to closing a deal. The carrier is usually in a position of power. The vendor usually must have fulfilled a slew of carrier needs before signing the contract. In other words, the vendor is forced to jump through hoops. Even then, the carrier may decide to "pull the plug" on the relationship at the eleventh hour.

The vendor must have its needs met in a contract. From the perspective of the vendor, the contract's terms and conditions should address the following issues:

- Vendor costs plus reasonable profit must be covered
- Vendor warranty must be capable of being fulfilled
- Contract time period must be acceptable to the vendor
- Delivery, testing, installation, and acceptance dates must be acceptable to the vendor
- The vendor must be able to fill the order as specified in the contract
- The vendor must be able to support testing as specified in the contract
- The vendor must be able to support ongoing maintenance
- Payment schedules must be acceptable to the vendor. The vendor should get some of the contract money upon signing, although a smart carrier will never pay for all the equipment up front.

The vendor will need a competent CFO, VP of Sales, and General Counsel to ensure that the company's needs are met. Each of the management team needs to respect and be able to work with the others. The last thing a vendor needs is a General Counsel who believes he can make the sale, a CFO who demands the first customer he ever signs pay for the vendor's past development costs, and a VP of Sales who wants to do the CFO's or General Counsel's job. This is a formula for disaster. However, for a carrier it usually is a good situation because it results in uncoordinated sales efforts and a desperate vendor.

Closing the deal is the goal of the vendor.

VENDOR ORGANIZATION REQUIREMENTS

Creating product is not a serial process; it requires a highly integrated and nimble management structure. Unless the company dominates a particular sector of telecommunications technology, it must have a highly focused and aggressive management team. Vendors in the telecommunications space (voice and data) need to be able to sense changes in the marketplaces of their customers and of their customers' customers. In a highly competitive marketplace, there is absolutely no rest for the management team.

Software, hardware, and services vendors share one common problem: the need to predict changes in the marketplace. Vendors need to be able to establish strategy and tactics for the company. They need an organizational structure that facilitates creativity and internal communications while simultaneously watching and listening to the marketplace. The organizational structure must be highly disciplined in its execution and flexible in its creative process.

The vendor needs to have a very strong (competent) marketing department capable of developing on-target market-driven strategies. A very large vendor may be big enough to make a mistake that will go unnoticed by the customer and investors. This is called "decision by error" and "flying by the seat of your pants." However, small vendors cannot afford decision experiments of this type. They need to focus on their strategies and tactics. The same can be said of small and large carriers. Key factors ensuring vendor success in the marketplace are:

- Customer focus
- Market focus

- The existence of a market for the product or type of product being sold
- Product that brings value to the customer
- Management team competence
- Management team integrity
- Consistency in high quality, service, and product
- Business strategy

This can be boiled down to three fundamental ideas: integrity, value, and vision.

The list of factors above is key to the success of a vendor. Figure 4.20 illustrates the core factors for success.

We can make a case that market dominance will always ensure success. However, more often than we might imagine, the sleeping giant vendor has been taken by surprise by the smaller desperate vendor. The successful small and large ven-

FIGURE 4.20 Factors for success.

dors have always focused on these factors first and foremost to establish the foundation of the company.

SUMMARY

Management structures for vendors are designed to optimize the vendors' creation, development, sales, and manufacturing processes. One of the keys to ensuring the success of the vendor is having a "game plan." This "game plan" is essentially the business strategy of the company.

A business strategy is a high-level statement of intent and of direction for the company. A component of the business strategy is the planning process, which is focused on the market.

The remaining chapters focus on market-based planning.

CHAPTER FIVE

STRATEGY, TACTICS, AND SHARED VALUES

Strategy is the art of defining how a person, army, or company intends to wage war. It points a company in a particular way but does not define turn-by-turn driving directions. Strategic planning is both science and art. Effective strategies are short, quickly articulated, and easy to understand. I often say that if you can articulate your strategy in five minutes and the listener just as quickly understands the strategy, then you have a strategy that people can act on. The fact that a person can understand a strategy quickly does not mean the strategy is any good; it just means that the strategy has been articulated well, which is an accomplishment in itself. Strategy does not define what the company provides as a product or service, but how the company intends to achieve its goal. *Tactics* define the actions needed to achieve the goals.

However, for any strategy to be implemented, it must be easily understood. Many telecom (and other business) professionals will too often become overly involved with the sound and tenor of words. These people want the strategy to have a high level of sophistication. In reality, bankers, financiers, and corporate executives want the strategy articulated in plain language. Anything else will be perceived to be hiding behind language and words. Early in my career, a financier said, "If you want the funding, be clear and don't waste my time. You have 5 minutes." Figure 5.1 illustrates the concept of strategy.

Strategies must be direct. Ideas must be clear in a strategy; they cannot contain subtleties. Company staff needs to be able to execute the strategy—therefore, the strategy needs to be crystal clear and without complexity.

STRATEGIC PLANNING

Strategic planning is a necessity in any company. In the world of telecommunications (voice, data, Internet, etc.) it is a fairly new concept. During the 1980s, all national telecom systems were operating monopolies. The amount of strategy that went into most of these companies was minimal by today's standards. However, global privatization and deregulation of telecommunications have left the industry struggling to learn simultaneously how to create a strategy and how to execute it in one of the most dynamic business sectors in the world.

Carriers and vendors approach the marketplace in the same way. The first thing is to scope out the marketplace.

Strategic planning is a necessity in any company.

- Strategic planning is a very dynamic process
- Strategy looks to the future

FIGURE 5.1 Strategy: What is it?

What is it? In other words, where do we play? How big is the market? What kinds of customers comprise the target market? What other companies are in the space we want to compete in? What value is the carrier or vendor bringing to the market? As many customers would say: "What do you have to offer? How is it going to help me? What is so special about your product?" The thoughts that go through the minds of the carrier and vendor are: "What is my value? What is my story? What bait can I dangle to entice this company or person to buy from me? Is the value obvious? Is the value I perceive real or just my believing my company's hype?" Every sale that is made, must have instantaneous and easily understood value or you will be perceived to have no product worth looking at, let alone purchasing.

Another aspect of strategic planning involves distribution channels. The carrier and vendor need a way of telling their story (the sales pitch), a place to tell their story, and people to tell their story to. This is a communications distribution channel. The next thing to think about is how to reach the market with your product. If you are a carrier, how are you going to get your service to the customer? If you are a vendor, operating in the United States but selling your product in Sweden, how do you get your product delivered? In such a case, a vendor may need an agent. This is the product distribution channel. The distribution aspect is far more complicated than I have described. I have just given you a taste of what is needed.

After covering these aspects of the strategic plan, a company must consider the scope and scale of the execution effort, for reasons of cost. The cost involves:

- The number of people the company needs to hire
- Salary levels of these people
- Commission plans of the salespeople
- The cost of building facilities
- The cost of manufacturing plant
- The cost of equipment

- Skill sets of the employees–kinds of skill sets and numbers of people with these skills
- Time frames for the execution effort

This list of activities is where the "rubber meets the road." It addresses one aspect of running any business: cost. How much is this endeavor going to cost stockholders, investors, and the company? How much and many resources need to be employed to get the job done? How much time will elapse before product is ready? It takes money to make money. As I have said in my previous books, telecommunications is a capital-intensive business requiring people and time. If you do not want to spend the money, then don't bother playing in telecommunications.

It might seem that price would be the last element of the execution effort. It is not price, but overall cost of the effort that forces a company to make a "go or no go" decision on a project. Worry about pricing once your company has decided there is an opportunity. Product price is ultimately dictated by the customer base. If the effort appears achievable only after great resource expenditure over a long period of time, then the company and its investors will vote against the effort.

We can easily categorize these elements of a business strategy. We need to know:

- Market segment and size
- Value—the story (the customer pitch that describes and differentiates the carrier or vendor from others)
- Distribution—how to get the company story told; how to deliver the product
- Scope and scale of execution

In fact, the elements of business strategy are applicable to any situation in which an individual or group of individuals is competing against others. There is an objective, and you need a plan to get what you want. Figure 5.2 is an illustration of the elements of a business strategy.

Strategy is about asking questions and making decisions based on the information available.

The company needs to know the following in order to create a good strategy:

- Market segment and size
- Value
- Distribution
- Scope and Scale

FIGURE 5.2 Elements of strategy.

Strategy is about asking questions and making decisions based on available information. The carrier and vendor need to develop a variety of scenarios in which a series of "what if" questions are asked and a series of possible answers are created. Strategic planning is a dynamic process. There is a goal, but the plan to get to the goal may need to change with time and circumstances. Strategy is not static. Strategy does not result in changing the product but it does change how to sell the product. Small telecommunications vendors must be especially good at making quick stops and turns in their strategies. At one point in the U.S. telecommunications technology industry, there were nearly a hundred different vendors who claimed to have a "switch-based voice mail product." The number was staggering because there were only two dozen major wireline carriers who were large enough to spend money on this product. During that time vendors closed their deals with carriers using price as the differentiator. Fifteen years later, the voice mail market is no longer growing and is occupied by a handful of companies. Those voice mail vendors who won out did so by evolving the product (in order to differentiate themselves) to include not only voice mail but also a slew of other capabilities that use voice mail as the basis of ancillary services. Vendors who could not anticipate and react to changes in the marketplace closed their doors.

The victory for many of these voice mail companies was a short-lived profitable one because the market suddenly changed. Today, carriers sell voice mail as a customer giveaway. The voice mail product is now expected to be included in most customer offerings. Further, answering machines for the home improved in quality and dropped drastically in price. Therefore, the original premise that voice mail was a differentiator no longer holds true. The market changed. Today, those wishing to make money in telecommunications technology consider the voice mail market a dead end for high profit margins.

The telecom technology business can be characterized as comprised of hundreds of early entrants attempting to be the "last person standing," with a resulting field of a handful of winners. The only way to win is by having a flexible strategy. Figure 5.3 describes the need for flexibility in strategy.

The elements of strategy are not disconnected. In fact, they are intertwined; each one affects the others. Strategic planning is an interactive process that needs to be attuned to the marketplace and the company's activities. The following sections will delve deeper into the elements of strategy in telecommunications.

MARKET SEGMENT AND SIZE

Whether you are a carrier or vendor, you must first decide where you want to compete. The company's management must identify the exact market it wishes to sell to. Our operating

Elements of Strategy

- Market segment and size
- Value
- Distribution
- Scope and scale

The elements of strategy are intertwined; one affects the other.

Strategy must be flexible.

Flexibility is achieved in the way the above elements are defined and redefined.

FIGURE 5.3 Flexibility in strategy.

assumption is that there is a specific product or service you are selling. The following questions should be answered:

- Is there a marketplace need that is crying out to be filled?
- Is the business space occupied by other providers?
- If the space is occupied by other competitors, then how many are in the marketplace?
- How big is the customer base?
- How big a piece of the customer base do I think I can capture?
- Do I know the customer base? In other words, do I have an understanding of what motivates the customer base?
- I know what customer groups I want to sell to, but are there other potential groups of customers I may reach? In other words, do I have a secondary marketplace to sell to?
- Is the market I am selling to a niche? An example would be customers who are in the domestic travel industry. Another example would be carriers who sell services to small businesses in rural markets.
- Will a niche business be successful?
- If I am in the niche market, is the market segment big enough for me to generate a profit and sustain my business?

The who, what, when, where, and why questions about the market segment are illustrated in Figure 5.4.

In the telecommunications business space, one can easily say that everyone needs to have telecommunications needs filled. However, the question that must be answered is one layer deeper. Is there a specific telecommunications need that should be filled but is not? If the need is being filled, then can it be filled differently, with better quality, and more cheaply? Does the marketplace even see that there is a need to be filled? This last question leads us into the value element of strategy, which will be considered later in this chapter.

Let us assume that we have identified a need and that we have a product value. The question of the existence of other providers of the service or product must be answered. If there are no

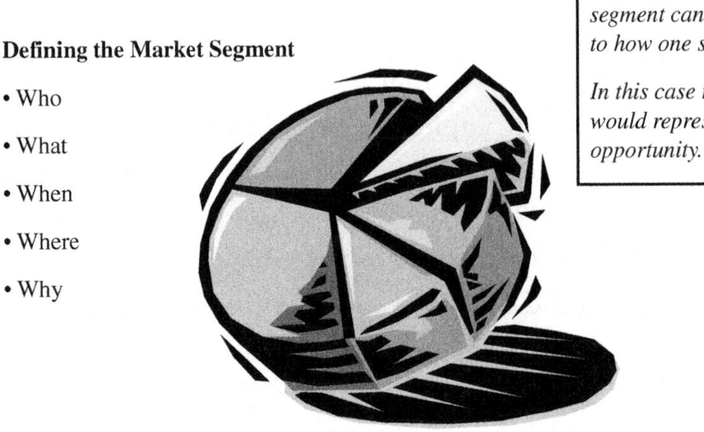

FIGURE 5.4 The market segment: who, what, when, where, and why.

providers of the service or product, then why not? Is it because we are the first, or is it because our belief that there is a need is nothing more than our own vanity speaking out? Now is the time to test our idea(s). We will need to dig deeper into the value element. However, let us assume there are other competitors in the market segment we wish to sell to. How many competitors are there? This is an important question because it will determine the intensity and skill set of our sales and marketing teams. The question will also lead us to examine the issue of value. If there are other competitors, what will my service or product do differently to entice the customer to do business with me?

The size of the customer base is important because it determines whether the telecom company will have a large enough number of customers to generate a net profit. If the customer base yearning for the product or service is huge, then do we want to sell to all of them, or only to a portion of them? A telecommunications carrier and vendor need to be able to look at an opportunity and assess whether or not the company has the resources to meet the demands of a large customer base. This leads into the issue of "scope and scale" of effort.

Knowing the customer base is a necessity. Even the small retail store owner selling meat and vegetables understands that you have to have on, the shelves, products the community wants. Many years ago, the dominant telecom sales approach was that a company could sell the customer what it had, not necessarily what the customer wanted. It was an old technique of sales intimidation that salespeople used to scare customers into believing a competitor would use this new service or product against them. This allowed many salespeople and companies to avoid spending time or money on doing significant market research. Unfortunately for the customer (carrier and user), the technique worked far too often. However, today carriers and users are far more informed and more skilled at negotiating. It is helpful to the customer, whether carrier or user, that there are so many choices for products and services. Knowing the customer base and what motivates it is a necessity. As I noted earlier, the motivation may be a history of doing business with the company, value of the product, or other factors. Motivation is a combination of fact (need) and emotion (preference based on subjective value points). Figure 5.5 illustrates these facts about the customer base.

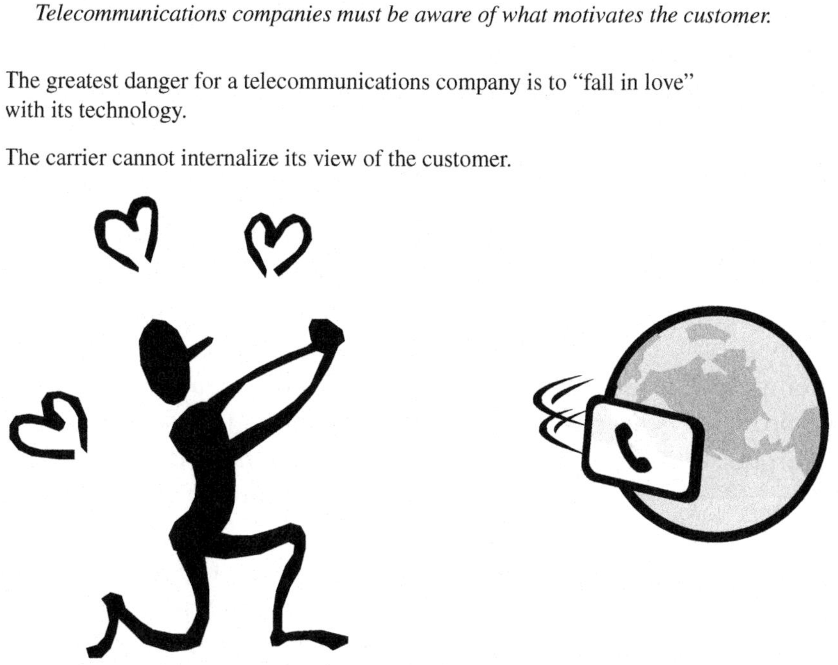

Telecommunications companies must be aware of what motivates the customer.

The greatest danger for a telecommunications company is to "fall in love" with its technology.

The carrier cannot internalize its view of the customer.

FIGURE 5.5 The customer base: motivation.

The size of the segment will dictate how much money the company makes. This is actually the last question investors will ask, and the most important one to have answered. It is also the last question a company asks itself, before it embarks on creating a product, and the most important one to have answered. It may become possible to expand the target segment to make money, but if we do that then we go back to square one and repeat the whole questioning process about the market segment. The result may be major or minor modifications to the product. The bottom line is net profit.

By now, we have begun to understand how this element of strategy is intertwined with other elements such as value. What we have seen is an example of how integrated the elements of strategy are. The following sections will continue to highlight this point.

VALUE: THE STORY

What is the value proposition? Why should the customer buy from us? As executives of our company, why do we believe the customer should buy from us? Do we have a product or solution that can help the customer? Why should the customer buy from us and not the other company? What we address in this section is the value aspect of the product or solution.

The greatest danger for a telecom company is to "fall in love" with its technology. The telecommunications service business is characterized by constant reinvestment and enhancement of the network in order for companies to stay competitive. Therefore, telecom vendors are constantly producing new product and technology. Very often, a telecom service provider is so enraptured by its technological solution that it forgets that the customer does not care how the service is delivered or works. The customer merely wants affordable and quality service. Think about it; how many people really care about how an automobile works? Except for the enthusiasts, the majority of consumers do not. Look at advertising on television; it is about luxury and horsepower. Telecommunications companies must be aware of what motivates the customer.

What gets that customer salivating to do business with us? Competition between telecom companies is focused on customer motivation. Carriers are motivated to buy from vendors in the same way most consumers are motivated in buying an automobile. In general, carriers like to buy product using the following parameters as their guidelines for purchasing (the parameters are also illustrated in Figure 5.6):

- Bigger
- Better
- Faster
- Cheaper
- The only solution on the market
- Relationship

Bigger refers to switching systems, transmission systems, and router capacity, among other things. The carrier wants to get as much as possible for its money. The more the product can do for the same amount of money as the other company's product (which may be doing less) the better for the carrier.

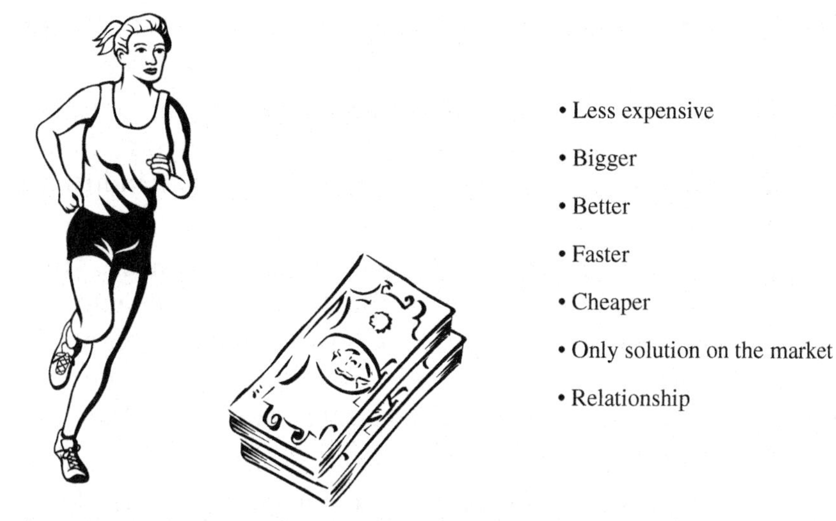

FIGURE 5.6 Motivations for carriers to buy from vendors.

Better refers to quality. Every carrier wants the highest quality of product without spending any more money than if the product were of lesser quality.

Faster refers to computing speed. Every carrier wants faster equipment. The faster the equipment, the faster information will be transmitted, and therefore the transaction, will be processed through the network.

Cheaper does not refer to quality, but to reduced expense. Carriers are always looking for ways of reducing their capital outlay. The term "most favored nation" does not refer to politics and a nation's status. The term "most favored nation" (MFN) is used to describe a carrier that is given preferential pricing from a vendor. MFN treatment is either the lowest price or the best contract terms; sometimes it is both. The MFN contract clause is sought by big carriers. Typically, the vendor will give a carrier MFN treatment for one or more of the following reasons:

- The carrier is the first ever to sign with the vendor.
- The carrier is placing the largest order the vendor has received to date.
- The carrier has verbally promised future work and the future contract could be very large.
- The carrier has agreed to be the test bed for a vendor's technology and now it expects to be given beneficial pricing treatment.
- The carrier has been a long-term customer of the vendor, even when the vendor had nothing really worth buying. Some vendors simply end up being active followers in the industry, always placing second in the field of innovation.
- The carrier threatens the vendor with loss of business. This succeeds because the carrier is the largest in the nation. Vendors who argue with this carrier deserve to lose their business. The first thing taught to retail salespeople is not to argue with the customer. Vendor salespeople need to be taught the same thing.
- Some vendors are so desperate for someone to buy their equipment that they give the carrier the best deal even if it

means losing money. Then the vendors hope to make up the difference on the back end of the contract through maintenance upgrades, new software, and network upgrades.

The carriers believe that they deserve to be treated well by vendors for a variety of reasons, which usually mirror the vendors' reasons just outlined.

The carrier–vendor relationship tends to be one-sided and balanced in favor of the carrier. This only works when the carrier is large. Small carriers tend to be at the mercy of the vendor.

Being the only solution on the market is still possible even in today's telecom environment. Many years ago, there was only one manufacturer for a device called the Signal Transfer Point (STP). The STP was originally the size of a 24-person conference room. The market was big enough for one vendor but not big enough for two. This is a guaranteed sale with the vendor in the position of control. Time passed and one day a second vendor appeared with an STP that was 75 percent smaller and less expensive. Suddenly, the market exploded because those small carriers who could not afford the STP of the first vendor's product could suddenly afford to buy the STP from the second vendor. The big carriers took notice and wanted to buy the less-expensive product. Now there was competition.

The relationship between the carrier and vendor is a motivator, although this is a complicated factor. Once a relationship has been created between a carrier and vendor, there will be recurring transactions. Global economic times have worsened and relationship has become less of a factor. Vendor loyalty is no longer a powerful motivator.

VALUE: THE PRODUCT'S/SERVICE'S INTRINSIC VALUE

Intrinsic value refers to what the product or service delivers to the customer. This does not mean pricing, contract terms, or even the intangibles like the good working relationship between the carrier and vendor, but what is being delivered in terms of specific product or service value.

Bigger, better, faster, and cheaper are the factors that lie at the heart of competitive positioning. Something has to be delivered to the customer and cost may not be a primary factor. What motivates the customer will usually be a combination of the factors listed. It is possible the customer will be motivated by a long-term relationship with the vendor. Typically, once the vendor has made a large sale, the carrier will choose to stay with that vendor because of the complexity and high cost of removing an infrastructure vendor's equipment from the network. However, in today's telecommunication market, with so many vendors providing the same type of product and carriers providing the same type of service, it is rare to see intrinsic value playing a dominating role in the purchase decision.

Intrinsic value is so difficult to use as a competitive positioning factor that many vendors do not even bother employing it as a differentiating tool. Other things being equal, the carriers will use price as the deciding factor. A vendor will say its product performs the work better, faster, and differently. However, this only works for the first few months and then a competing vendor will have replicated the product using different technology and methodology (without violating the other vendor's patents). Product knowledge, as a competitive advantage, is difficult to control and easy to replicate using different methods. The carrier does not care about how the product does what its does.

The carrier encounters the same difficulties when dealing with the retail user, its customer. The end-user does not care how the service is delivered, just that it is delivered at a good price. The challenge to carriers is to find what motivates their customers. What motivates a carrier to do business with a vendor is not necessarily what motivates a user to do business with a carrier.

To the user, today, every carrier seems to offer the same voice, data, and Internet services. This is often because the carriers all buy from the same suppliers. There are only so many ways a carrier can offer voice mail, call forwarding, email, and Web surfing. The result is that the user ends up making a decision based on the following parameters:

- Pricing—the principal factor in a user's decision

- The ease of using the service
- How easy it is to understand the bill
- How easy it is to understand the multitude of rate plans (the fewer the better)
- Professionalism of the customer care staff

The value that the carrier perceives in a particular vendor's product usually does not translate into easily understood value for the consumer. Many carrier technologists forget that the fact that a particular database can process queries 20 percent faster than another vendor's database is meaningless to the retail user. The carrier must create a value proposition for its users. Figure 5.7 is an illustration of how differently these several parties are motivated.

VALUE: THE VALUE PROPOSITION

The value proposition is part fact, part half-truths, and part art. How the sales pitch is created is part science and part art. The pitch is the story the vendor wants to tell. It must be interesting and relevant to the customer. The customer must want to hear the story. The story must end with why the customer wants to do business with you. The "why" is the value proposition. A

- Customers are motivated by different things; sometimes it is a combination of two or more factors.
- Customer motivations will change over time.
- Cost and Performance tend to be the dominant motivators across all customer groups.

Customer motivations need to be understood.

Carriers and vendors cannot effectively do business with their customers unless the customers' motivations are understood.

You cannot create a sales pitch or product unless you know what the customers needs.

FIGURE 5.7 Motivations.

telecommunications value proposition is an amalgamation of everything we have discussed. The pitch is a constantly changing story that is molded, then pitched to the customer, and then remolded based on customer input. The value proposition only stabilizes when the correct tenor and tone have been reached.

Once the value proposition has been identified, the vendor needs to be aware of what its competitors are doing. Competitors are modifying their products to create new kinds of value and lowering their costs to keep prices low, thus the first vendor is constantly forced to seek a competitive advantage, which means changing the value proposition.

Vendors and carriers will sometimes make the mistake of focusing only on the future. Instead, they should spend time looking over their shoulders at the competitors nipping at their heels.

DISTRIBUTION

This is usually the last thing manufacturers and carriers think of. Until the late 1990s, carriers viewed sales distribution as an aspect of the business they controlled. Prior to the Telecommunications Act of 1996, which eventually set off a firestorm of global activity, there were only a handful of carriers. The 1996 Act created hundreds of new carriers. This caused a surge in new vendors (with new technology ideas and products) and made existing vendors expand their operations. When there was only a handful of carriers, sales distribution and product distribution were not considered major issues. Prior to 1996 there was only a handful of customers and suppliers.

The resulting large number of carriers caused the traditional telecom vendors to expand operations to meet demand and to create new organizations to come up with new product ideas. New vendors were appearing overnight to create and sell new products. The carriers and vendors needed to pay attention to how distribution channels affected their businesses. Distribution addresses the following issues (see Figure 5.8):

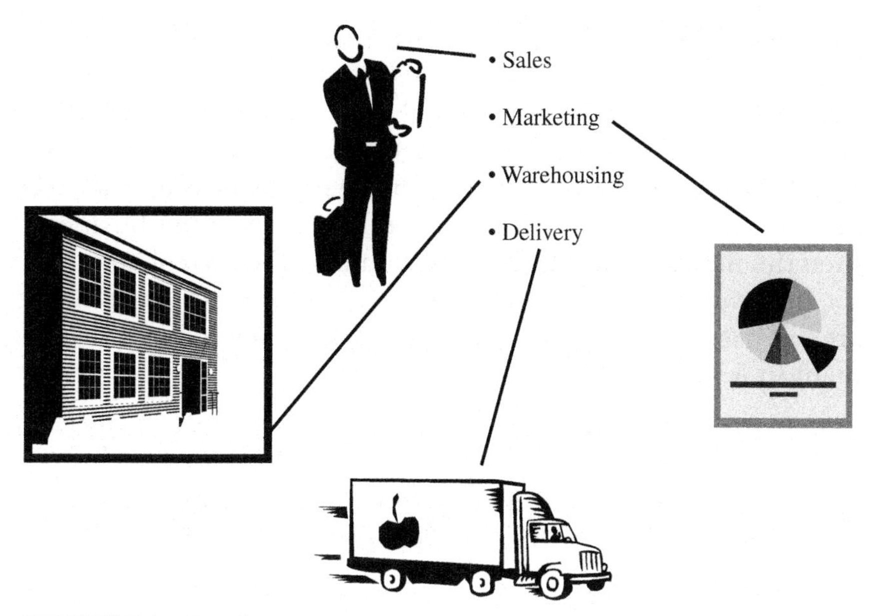

FIGURE 5.8 Distribution.

- *Sales.* The number of salespeople, the location of the salespeople, the structure of the sales force, points of sale and contract representatives (agents)
- *Marketing the product.* Advertising and methods of marketing the product or service
- *Delivery.* How the product is delivered, who is delivering the product, and delivery processes
- *Warehousing*

DISTRIBUTION: SALES

From a carrier's perspective, warehousing is the only area of distribution that is not a problem. Carriers sell services; therefore, a huge area of concern is the way in which the services reach the users. This is both a network and device issue. Wireless carriers must be able to get the device into the hands of users. They must have stores in high-traffic areas in order to reach the maximum number of users, and this naturally leads to the next concern: the number of salespeople to hire, sales training, customer sales incentives, and commission plans. From a wireline perspective, the user can go to any retail store and pick up an inexpensive telephone. However, the wireline carrier needs to ensure that the network is properly engineered to deliver the service to the user's place of business or home. If the carrier were in the Internet service provisioning business, then the network, not the device, would be the area of concern.

In the case of the telecom infrastructure vendor, storefronts are not used to sell equipment to a carrier. A vendor sells equipment in a variety of ways:

- Directly to the carrier via its own sales force
- To a carrier via an OEM relationship with a bigger vendor who happens already to be selling to the same target carrier
- Via another carrier
- Via agents

Figure 5.9 is an illustration of the paths of selling equipment.

Selling via another carrier is a fairly new phenomenon. The carrier effectively represents the vendor in a relationship with other carriers. The carrier will make enough money on the deal to cover costs. The reason a carrier can represent the vendor in

- Carriers – agent sales
- Agents
- Direct Sales
- Third Parties
- OEM Relationships

Multiple avenues for selling equipment can be taken.

The important thing to remember is maintaining account control.

FIGURE 5.9 Telecom equipment: avenues of sales.

sales to another carrier is because the carrier is selling its maintenance and operational expertise.

Selling equipment via agents is usually done when a vendor has no native presence in another country. In order to gain an immediate international presence, the vendor cuts a deal with key local business people. These people know the local customers and unwritten rules of doing business in their locale. They have personal relationships with key telecom carrier executives. These agents are paid by the vendor to represent the product. They can be described as *sales mercenaries*. The agents typically are paid a percentage of the sale, and usually more than the commission paid to salespeople the company employs full time. This method is used when a carrier is not capable of fielding a sales force in a country in which it does business.

DISTRIBUTION: MARKETING THE PRODUCT

Not often discussed is how the vendors and carriers promote their products and services. Carriers promote their product by using a number of different forms of media: television, radio, Internet, billboards, and print (newspapers and magazines). Getting the product advertised is part of the distribution model of the product. What is being distributed is information about the product.

Major equipment vendors and carriers do not rely on advertising in print media to communicate their products' capabilities. Rather, the media is used to hone or fine tune the image of the product. The image that the carrier sees is created by the product's performance, past performance of the vendor, and one-on-one discussions between vendor and carrier.

This part of the distribution chain is managed by sales and marketing. A toolkit for salespeople must be created that contains all the collateral sales material about the product and service. The marketing staff identifies the target markets and provides the salespeople with general market information. The marketing department sets the direction the salespeople follow, and the salespeople execute the strategy.

The marketing department is responsible for setting the message and image of the carrier. Working with public relations, it finds ways of communicating with the user. The public relations people usually have all the direct and close contacts with various forms of media. As I noted before, the carrier uses a variety of media to get its message across. It may even use movie stars to advertise its product. Users may like the movie star making the sales pitch; they enjoy hearing from celebrities they know. Users like to watch entertainment showing their favorite movie stars and associate quality with famous names. Users like to buy from people they like.

A major equipment vendor (of fiber optics, switches, routers, etc.) gains very little from advertising in the media. The media speaks to the user community, which happens to be the customers of the carriers. If the vendor is a terminal device vendor, the media advertising approach will work. Major equipment and terminal device vendors promote their products at a number of annual industry trade shows. These are the times when vendors show off product and issue press releases. The trade shows are also times when carriers can meet with many vendors during the week, undisturbed. Vendors show off their companies by taking targeted or existing customers to dinner and entertainment.

Figure 5.10 is a depiction of the avenues of sales and marketing communications employed by carriers and vendors. It illustrates the relationship among the various communications paths between the carrier, the vendor, and the user.

Telecommunications trade shows are annual events that enable vendors to speak to customers whom they had not originally considered. These carriers will "walk the (convention) floor" looking for new products. Carriers and vendors use trade shows only to meet with one another but also to make company announcements in press releases distributed at the show or during a trade show–hosted seminar. The entire industry is gathered together for a one-week period, talking about the latest industry developments; carriers and vendors would be foolish not to take advantage of the event. A trade show can also be described as an event with a captive audience. Media events, like trade shows, cannot be ignored by vendors.

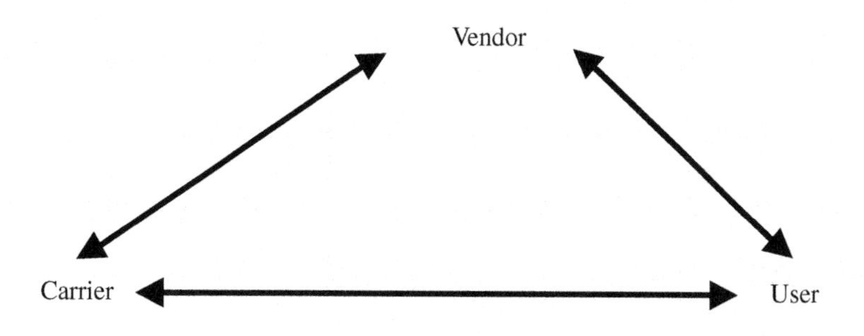

Communications between the suppliers and the customers can be likened to a circular relationship.

FIGURE 5.10 Communicating between the supplier and the customer.

Getting information out to customers is a daunting task. A vendor has a limited amount of time and money to get information about its product distributed to carriers. The vendor needs to rely on word of mouth, trade shows, face-to-face meetings, and technology trials to reach the carrier community. The carrier needs to reach its customer base as efficiently as possible, using those forms of communications customers tend to use for entertainment and information. The carrier needs to get its message out in 30-second bites of time for television and half-page ads in the newspaper.

The technology trial has an interesting place in the information dissemination aspect of the distribution phase. Carriers will not deploy any product unless the product has been tried out in its network. The carrier community will test technology being developed and new products in the network.

Note that products are technology that has been thoroughly tested and made ready for commercial deployment. A carrier considers a deployment to have commercial status when the product is serving paying customers. Trying out a technology in its infancy serves the vendor in the following ways, by:

- Creating a strong bond between itself and the carrier
- Enabling the vendor to see how the technology would work in a commercial environment

- Providing insight on how the carrier might wish to use the technology to provide a service
- Providing network operations field insight from the carrier: how big the product should be, what kind of performance metrics the carrier will be seeking, etc.
- Providing insight on carrier marketing initiatives, which will end up affecting the product's capabilities
- Affording time to begin the process of testing the carrier's price point to buy the technology when it becomes a product

Performing trials of an existing product provides the same kind of information. However, in this instance information gathered about product enhancements will be used in the product's next iteration. The data gathered from the carrier are still immensely helpful.

The trial supports the sales process and the vendor knows this, which is why an account manager is assigned to the technology or product trial from Day One. It is understood that if the trial is not managed well vis-à-vis a professionalism–carrier relationship perspective, the vendor will lose the sale. If the trial goes extremely well, part of the sales effort has been completed. Figure 5.11 is an illustration of how the technology or product trial fits into the customer communications process.

- Creates a strong bond between the vendor and carrier
- Test new technology
- Vendor gains insight into the customer
- Vendor sees how the product will behave in a live environment
- An opportunity to develop new concepts based on the results of the trial
- Negotiating price

How the vendor responds to trial problems and to the carrier's questions will ultimately decide the fate of the vendor.

FIGURE 5.11 Technology/product trials.

The carrier will always expect a technology trial to have moments of crisis. How the vendor responds to trial problems and to the carrier's questions will ultimately decide the fate of the vendor. No one likes to do business with a vendor perceived to behave either unprofessionally or incompetently during a crisis. Remember that if the product has operational problems in a commercial environment, the carrier usually does not perform equipment fixes—the vendor does. Carrier executives will not gamble with their careers on a vendor who cannot perform in a crisis.

The technology or product trial is such an important sales tool that vendors beg to have their technology or products tested by carriers.

DISTRIBUTION: WAREHOUSING AND DELIVERING THE PRODUCT

For vendors, managing inventory is the difference between financial health and ruin. Hardware manufacturers need to build systems to order. Hardware vendors cannot build switching systems, routers, or optical electronics for the purpose of storing the equipment just in case a carrier places an order. Hardware is money. Hardware in a warehouse is money sitting around doing nothing for the vendor. Vendors essentially build equipment for signed orders.

Hardware vendors need to project sales for the year ahead in order to schedule all the appropriate resources needed for the vendor to fill a customer's order. The following activities will take place in a hardware vendor's shop in order to prepare the company to meet customer needs:

- Projecting orders for the coming year
- Scheduling when the projected customers will need the units
- Alerting suppliers that there may be one or more big contracts signed, so that they can get ready for a large order of components
- Making sure to have enough people and the right people on the payroll to perform the installations

- Making the sale
- Getting a signed contract
- Getting a schedule for delivery and installation

Vendors make projections and schedules. Hardware vendors work hard at scheduling their own resources and the resources of their suppliers. They cannot afford to warehouse tons of equipment. Vendors may keep spare parts around, but not whole routing systems or 1,000 kilometers of fiber optic transmission facilities. If a hardware vendor kept that much equipment sitting in a warehouse without a customer name (and contract) associated with each piece of equipment, the vendor would be out of business. A vendor warehouses a minimum number of components for maintenance. Whole systems are warehoused only because the signed customer is awaiting delivery. Carriers are aware that vendors manufacture upon a signed order and will place two kinds of orders: a fixed number of large systems ordered (up front at time of contract signing) over a schedule to be agreed upon at a future time, or orders that stream in over a period of time. The carrier will demand some kind of proof that the vendor will be able to meet its needs when an order is placed.

The carrier will specifically seek proof that the vendor will be able to meet its time schedule, although proof is a difficult thing to obtain. A carrier will typically ask for the names of the vendor's suppliers, a tour of the vendor's facility, a meeting with the executive team and the senior staff, and a trial of the product. The trial is a way for the carrier to see how fast and how well the vendor's teams of engineers and technicians react in the field. The trial is usually large enough to force a vendor to spend a great deal of money, without being compensated for the trial equipment. Seeing how a vendor responds to hearing a six-month trial is going to be conducted at the vendor's expense is a good test of vendor resources, commitment, and capabilities to manufacture and deploy. Carriers are in a position of control.

The whole delivery process for telecommunications equipment is much the same as in any industry. Hardware vendors will often use their own subcontractors to deliver equipment to customer sites. These types of subcontractors are experienced in the handling of sensitive computing equipment.

Carriers may deploy a new service either at once or over an extended period of time. The first situation is a "flashcut" and the other is a phased-in process. The carrier needs to have the technology installed over a large coverage area and the service needs to be accessible to all customers. Carriers do not manufacture, they essentially distribute a service. The distribution of the service is via the network a carrier operates. The network needs to be able to provide the service within the same quality-of-service parameters for all customers and in a cost-effective manner.

From a strategy perspective, the whole discussion of distribution is important because distribution is a parameter of execution. If the initial discussions on distribution indicate a high level of difficulty in filling a carrier's order or ensuring the service can be provided to users, the company may decide not to manufacture the equipment or sell the service. Figure 5.12 illustrates the components of manufacturing and delivery.

SCOPE AND SCALE OF EXECUTION

The scope and scale of executing an effort is a component of other strategy elements. The question of "how big the job is" is the ultimate question that needs to be answered before a company embarks on any endeavor. From a strategy perspective, the scope and scale of an endeavor may be so large and take so much time to achieve that the vendor or carrier may not pursue a particular path to achieve its goals.

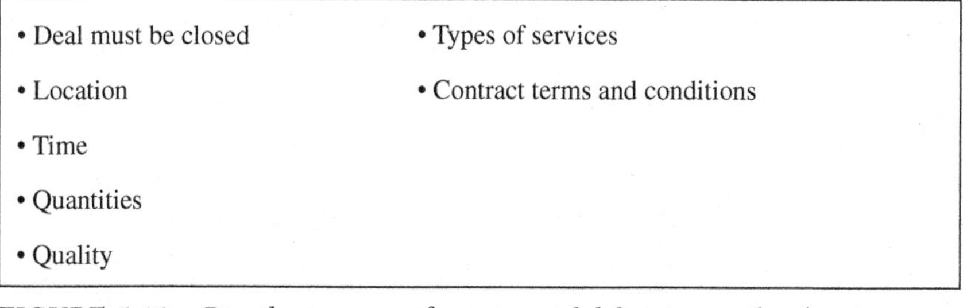

FIGURE 5.12 Distribution: manufacturing and delivering product/services.

The carrier may decide to wait until the product is less expensive or can be more easily operated. Or it may decide the customer education aspect of providing a service is so complicated that customers would find the service unusable and not buy the service. The vendor may see that the time frames associated with entering a particular telecommunications segment are far too long for the vendor's financial objectives and therefore abandon the product. The vendor may have a specific market size objective or t may see the market already flooded with other vendors. In order to achieve market dominance, the vendor will need to expend enormous amounts of money, time, and human resources in selling the product. If a vendor chooses to achieve market dominance for a product by undercutting competitors' pricing, the vendor needs to be confident that there will be recurring business over an extended period of time so that it can eventually meet its financial objectives.

Scope and scale of an endeavor addresses the following parameters: time, size, and cost.

Telecommunications is a business that operates in short market windows, though much time is required to deploy product. Telecommunications is a capital-intensive (money and equipment—software and hardware) business. Money is constantly being invested in the network. The overall size of every effort is large. A service is never sold to just one customer; it is usually sold to multiple customers. The only time a service is sold to one customer is when that customer is the national government, the largest single user of telecommunications.

TACTICAL PLANNING

After the company decides on a strategy, it needs to determine how to execute this strategy. This is called tactics. Tactical planning does not define or redefine strategy, but rather defines the step-by-step directions for how to achieve the goals of the company. Figure 5.13 illustrates the role of tactics.

A carrier's or vendor's tactical plan involves the following:

The vision defines what the company wants to be.
The strategy defines the specific set of goals needed to fulfill the vision.
Tactics defines the step-by-step actions of the company.

- Tactical Plans are executable
- Tactical Plans define actions

Examples of Actions:

Vendor selection

Technology selection

Service bureau selection

Selection of accounting firm

Enforcing financial controls

FIGURE 5.13 Tactical planning.

- Project schedules
- Vendor selections
- Staffing requirements
- Work assignments
- Creating a short list of target customers
- Creating a sales plan
- Creating an advertising campaign
- Defining the services
- Defining the product
- Establishing meetings with customers

Some call tactical planning *action planning* because it defines specific actions that need to be undertaken. However, something is needed to keep actions aligned with the strategy: this something is called *vision*.

BINDING STRATEGY AND TACTICS: THE VISION

A strategy establishes general directions or objectives. Tactics establish detailed step-by-step directions. The binding element is the vision. Every company has one. The vision is a rallying point for every member of the company. The company's vision binds strategy and tactics. An example of a vision for a carrier might be: "Our company will be the nation's premier provider of residential Internet services, bringing broadband services to the underserved markets. We will achieve this goal by working with the communities and treating them with respect." An example of a vision for a vendor might be: "Our company will be the nation's premier provider of children's educational software broadcast over the wireless Internet, focusing on children in grades K–5."

The vision defines what the company wants to be. The strategy defines the specific set of goals needed to fulfill the vision. The examples I have given may not be the most inspiring visions ever written, but visions are supposed to galvanize, empower, excite, drive, generate passion, and motivate employees to exceed their own expectations of themselves.

Visions create the shared values of the company. This is the first thing taught to company managers. The vision articulates the mission of the company. A corporate *mission statement* is a component of the vision. The carrier's mission may be to provide residential broadband services. The carrier's vision "kicks it up a notch" and articulates the carrier's goal of being the best and providing the service at the lowest prices; more important, the carrier wishes to achieve the goal of bringing these services to underserved markets. Every vision articulates a desire to serve the community it is targeting. Visions cannot state, "we want to make lots of money and have high stock prices." If a corporate vision actually stated it wanted to make lots of money, no one would want to do business with the company and employees would not want to work there. It is understood that the company wants to make money. Nearly all large companies have visions that articulate a desire to be the biggest and best. However, to say you want to be the biggest and best is not a necessary component. Companies are created for one reason: to sell something to customers at a profit.

Visions need to be written down and articulated to every employee with intensity, passion, and conviction, because the company needs to articulate the vision over and over to every employee using the same phraseology so that all understand the vision in the exact same manner.

Telecommunications companies are rushing to get products or services to market at a pace so fast that it is easy to lose focus. The vision binds together the strategy, tactics, and the employees. Vision defines what the company does for a living. Figure 5.14 illustrates the role of corporate vision. Vision dictates the makeup of the team. We will cover this in detail in Chapter 8.

WHERE DO VISION, STRATEGY, AND TACTICS LEAD THE COMPANY?

Vision, strategy, and tactics are components of running a company. The challenge for the telecom carrier and vendor is staying focused on the market. The vision may define the company, but the focus on execution can still be different. Execution

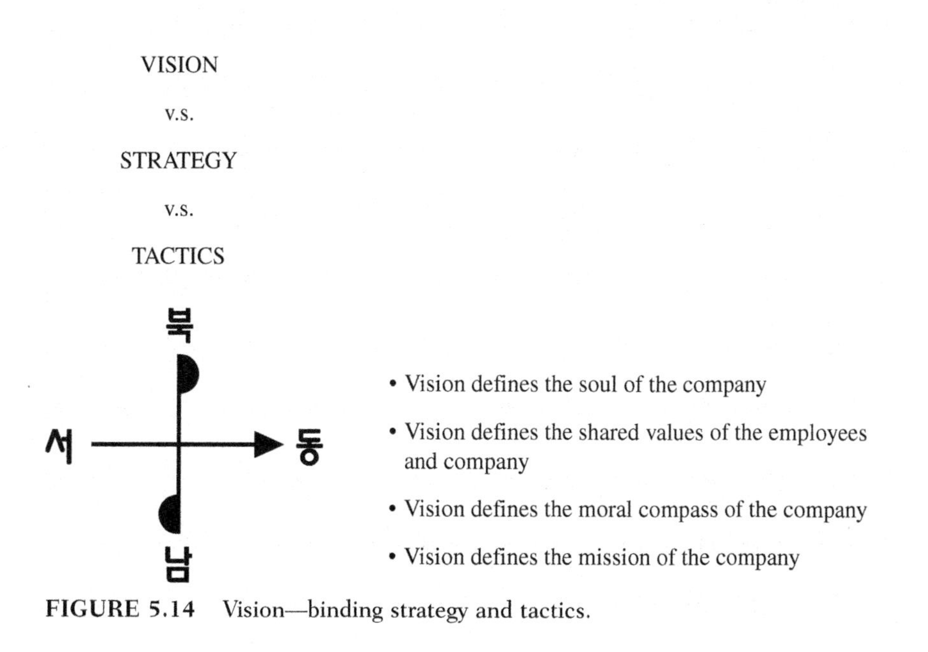

FIGURE 5.14 Vision—binding strategy and tactics.

focus can be lost even though the company has a vision, strategy, and tactical plan.

Telecommunications carriers and vendors are technology-focused companies. It is very easy to get lost and forget that both are working to serve customers. In the 1970s, it was a common sales and marketing tactic to make product and work hard to sell customers on the idea that they needed the product. The companies were not market driven but purely sales focused. Companies would tell customers that Product XYZ was all they had and nothing more; customers should not expect the vendors to do anything more. Telecommunications carriers were notorious for this attitude. In North America, the saying was, "you can have any color telephone you want as long as it is black." No attempt was made to understand what might make the customer want to buy more of the product. The focus was on sales and meeting one's "numbers." As important as meeting sales projection is, unless the company maintains a market focus, it runs the risk of getting too comfortable, because projections are being met while marketplace shifts are ignored. Figure 5.15 is a depiction of the need to maintain a market focus.

Today, the goal of a telecom carrier is to meet the needs of the customer. In a market environment filled with so much competition and change, maintaining a market focus is critical. We will address this issue further in Chapter 6.

Focusing on the customer, not focusing on just profit

Focusing on the needs of the customer, not focusing on meeting sales quotas

Focusing on needs

- Cannot focus on your own corporate hype
- Cannot focus on the network
- Cannot internalize your view
- Carrier must look outside of its business and network

FIGURE 5.15 Market focus versus sales focus.

SUMMARY

Vision, strategy, and tactics are critical components of running a business since they define the company, its goals, and how it will achieve those goals. Vision, strategy, and tactics also serve as components of a company's business plan. A business plan defines what the company is delivering to its customers and how much money the company expects to make over time.

Without a vision, there is no telecommunications company. Today's economics and dynamic market environment demand that a telecom company possess a realistic and achievable vision. Achieving balance between vision, strategy, and tactics is an ongoing challenge for any telecom company.

The telecom industry has been rocked by economic hard times. Achieving balance between vision, strategy, and tactics is difficult when return on investment is a goal. As the regulatory bodies and ILECs have known for generations, telecommunications is a part of the fabric of everyday life. The need for telecommunications will not go away. The growth and evolution of telecom technology is great and the market is undergoing constant, continuing change.

The one thing that a telecom professional can predict is change in the marketplace. Established telecom companies will continue to look over their shoulders, waiting for the start-up that has the right business plan, the right products or services, the right balance between financial goals and execution, and the right management team to dominate the marketplace. The marketplace is dynamic and the telecom marketplace is an excellent example of this environment.

The next chapter discusses staying on focus for a market-driven versus sales-driven company approach.

CHAPTER SIX

MARKET FOCUS—UNDERSTANDING THE TELECOMMUNICATIONS MARKET

The telecom marketplace can be viewed from two perspectives: vendor and service provider. The telecommunications service provider is a company that provides service (to the user). This type of telecom company is also generically called a carrier or a service provider. The user can be an individual or a company. A user purchases telecom service from a carrier/service provider. The different kinds of carriers are:

- Wireline telephone companies
- Cellular carriers
- PCS (Personal Communications Service) carriers
- Paging companies
- Cable television companies
- Satellite carriers
- Internet service providers

The vendor community provides software, hardware, and services to its own members and to service providers. As I noted in Chapter 1, hardware is a generic term referring to all the physical equipment made or assembled by the telecommunications company and can include switches, chips, routers, wire, insulation, digital signal processors, generators, batteries, and tools.

Software can be defined as instructions given to a computer or some piece of hardware. It is used by vendors and carriers to operate telecom equipment and provide services.

The overall telecom marketplace can be classified as retail and wholesale. The retail business space includes companies that sell directly to the principal users of their equipment and services. The wholesale business space includes companies that sell equipment or services through at least one other party. Figure 6.1 is an illustration of the retail and wholesale business space.

RETAIL TELECOMMUNICATIONS SERVICES MARKETPLACE

The retailing of telecom equipment and services is what most people refer to when speaking about the telecom business. Retail services space can be divided into the following areas:

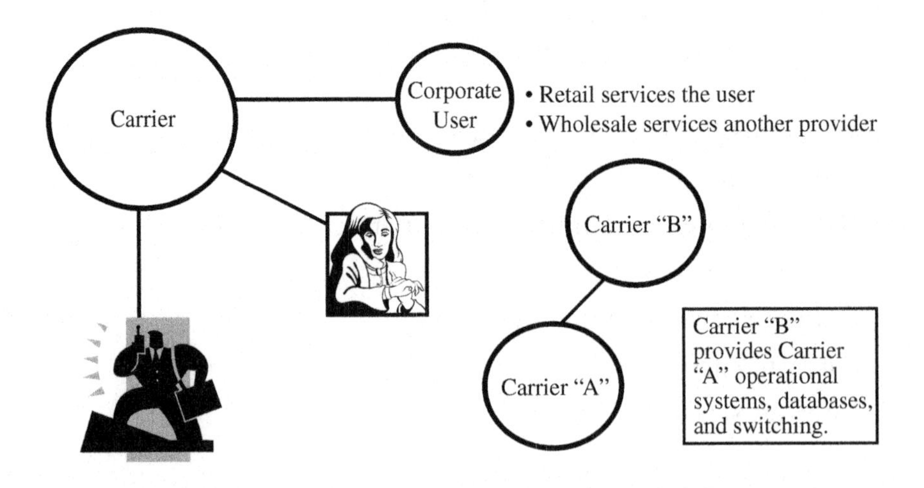

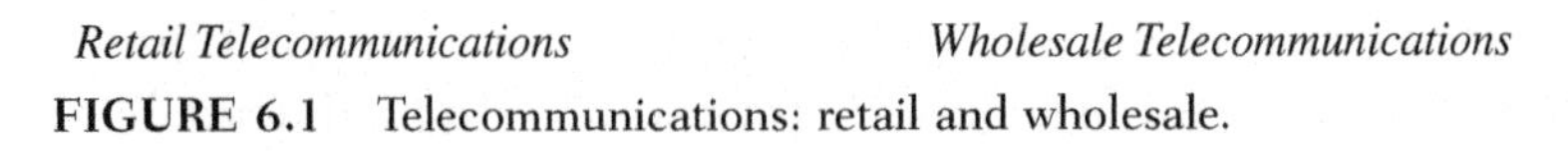

FIGURE 6.1 Telecommunications: retail and wholesale.

- Residential—apartment buildings/multiunit swellings
- Residential—private residences
- Small business
- Medium business
- Large business
- Government

Figure 6.2 is an illustration of the retail telecommunications space.

RESIDENTIAL

The residential space (single- and multifamily dwellings) has undergone a major transformation since the 1990s. Once, the only service brought to the home was Plain Old Telephone Service (POTS). POTS is basic wireline telephone voice service. The difference between today and the early 1990s is that residential users now have access to data services.

The residential business space is now comprised of voice services and data access services. The data access services I refer to are the Internet connections residential users are given access to. The data capable connections include:

- XDSL—X Digital Subscriber Line technology
- Cable television
- Satellite access

Selling to the User	Product
• Residential	• Voice
• Business	• Video
• Government	• Data
	• Broadband

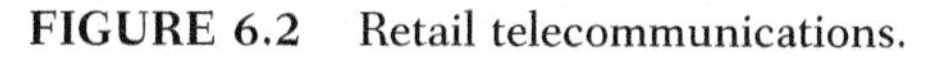

FIGURE 6.2 Retail telecommunications.

- ISDN—Integrated Services Digital Network
- FTTH—Fiber to the Home (on the near horizon via the passive optical network)

Services that will soon be available for purchase by residential users include:

- Video communications
- Music
- Video entertainment
- Online radio—via wireline instead of wireless connections

No specific service is listed because the problem of the current broadband residential market; there is no single *killer application*. The residential community is not an easy one for which to create value beyond the basic Internet connection and voice service.

Many residential users seek data rates on the orders of megabits per second, but they do not constitute the mass market. Some carriers are able to generate positive net profits in a niche business market. Unfortunately, small carriers run the risk of not being able to generate future value without a significant investment.

Segmenting the residential space has been a problem for most Competitive Local Exchange Carriers (CLECs), because there are few easily identifiable telecom needs other then voice and Internet access. The problem with voice is that it is no longer a high-margin revenue generator, and Internet access is usually accomplished via dialup. Segmenting the residential space should begin with understanding who occupies the typical residence. The demographic elements of the residential market that a carrier or equipment supplier should consider include:

- Average household size
- Average number of units in a multiunit dwelling
- Income levels of households
- Gender

- Employment categories—white collar, blue collar, doctor, lawyer, engineer, laborer, etc.
- Number of people in family working full-time jobs
- Number of children in household
- Age of children in household

These elements are easily understood by vendors selling equipment to a residential base. Take, for example, a vendor making wireless handsets. By allowing the user to choose the color of the device and change it by locking a colored faceplate over the plain one, a simple and ingenious way exists for the vendor to market thier device to a particular customer. The handset stands out among the typical black and gray models, and users are attracted to these handsets because, all things being equal, the multitude of handsets have the same level of quality. User-created handset color schemes make such handsets status symbols. The handsets' display can even be customized, so that users create their own display messages.

Someone running a sales and marketing organization(s) should be asking the following questions:

- Who is buying these handsets?
- Is it the 20–40 year-old age group?
- Is it the 30–45 year-old age group?
- Where are the bulk of the sales being made: big cities or rural areas?
- Are the buyers mostly female or male?
- Should the price of the faceplate be increased or lowered to stimulate handset sales?
- How does the faceplate actually play a role in increasing handset sales? What is the next step? Should you find another easy-to-execute handset change that will not cost the company large incremental investments in new technology?

Selling services can be much harder than selling equipment. At some point, every carrier faces the moment when a competitor is selling the same service. Carriers do not file for patents or intellectual property protection for new service ideas. Processes may be filed for patent protection, but this offers little competitive advantage. Customers/users really do not care if the carrier's technical process or technology enables it to provide the service more effectively than another carrier could; users care only that the service works and works inexpensively. If this is the case, why even bother segmenting the market? The answer is, because segmenting is a critical factor in gaining market position.

The goals of carriers are to get to market and to dominate their business space. Segmenting the market enables the carrier to provide and price the service to the customer as accurately as possible. Carriers rarely get a second chance to sell a service to users once they have failed to meet user expectations. The average customer is unforgiving when he is mistreated or overcharged. Understanding the segment is critical: the demographic elements described are examples of the level of marketplace understanding a carrier needs. Figure 6.3 is an illustration of the residential issues just discussed.

The revenue potential of the residential customer has yet to be realized.

The dominant service today is voice.

The growing service is Internet access.

The hope of most carriers is that broadband service to the home becomes the next VOICE service—the staple service.

FIGURE 6.3 Residential segment.

Another challenge for carriers is setting customer expectations. A carrier cannot say it will sell a service to a user and suddenly not provide that service in the expected manner. Carriers need to understand what their competitors are selling. For example, an ILEC provides voice and Internet access. This ILEC even has its own email, Web site, and search engine. The price for buying the Internet package is less than a current popular Internet Service Provider (ISP) charges. The ILEC will provide voice with the Internet package for just a small additional amount. The total package is still less than the ISP's email, Web site, and search engine-only service. Worse, the popular ISP can only be reached by dialup via the ILEC's network. However, the ISP's Web site is easier to navigate, much more user friendly, and much more colorful. The first-time users get the first month free, and long-time customers get six free rolls of 36-exposure 35mm film. Guess who wins: the ISP without the network. Customers do not buy services just because a carrier perceives its own product to be of higher quality.

Carriers need to conduct market surveys to determine what motivates a customer segment and to decide which customer segment they wish to target. A carrier needs to determine if the price it wishes to charge will be attractive to users. A key point to keep in mind is that if quality, ease of use, and customer service are perceived as equal, price will ultimately be the determining factor.

Elements of the service provisioning business are very difficult to differentiate. A carrier that deploys a service first needs to take the lead quickly. The customer will listen to the carrier that speaks first. Market advantages for a carrier in the retail space are fleeting. For example, one wireless carrier became the first to provide users with a one-rate plan across the nation. This plan enabled the carrier to capture a major portion of the customer base. Today, every carrier has a one-rate plan. The plan differentiated that carrier, but only for a short time.

Some may call this the "copycat approach to sales and marketing," but the customer doesn't care. The reality is that the customer is right and no one will care who is correct at the end of the quarter when financial results need to be reported. In other words, if it works and makes money, then it's a good idea. There will be more on market segmentation later in this chapter.

BUSINESS

The retail side of providing services to the business community is a highly competitive one. Most carriers initially target the large business customer. Many carriers may use the following numbers of voice lines as delineations for the small, medium, and large business customer:

- *Small.* Up to 500 lines
- *Medium.* 501 to 2,500 lines
- *Large.* 2,501 lines and beyond

The designations used to have a single common meaning to the old Bell system. The larger the number of lines of service being provided, the more important the customer. However, these designations are literally meaningless today for the following reasons:

- *Customers come in all sizes.* What is a small customer for one carrier may be a large customer to another carrier
- *Customers need data support.* Internet access requires data transmission facilities and system support. Data is a higher-profit margin service.
- *Customers have divided their telecommunications needs among multiple providers.* The pie of opportunity has gotten smaller and so have the slices of the business.

Figure 6.4 is an illustration of customer segmentation by size. Note that most designations are based on the number of voice lines, because this gives an indication of the size of the customer.

The thinking in the old Bell system was that if you had the voice business, you would obtain the rest of the customer's telecom business. This is a key point to bear in mind because

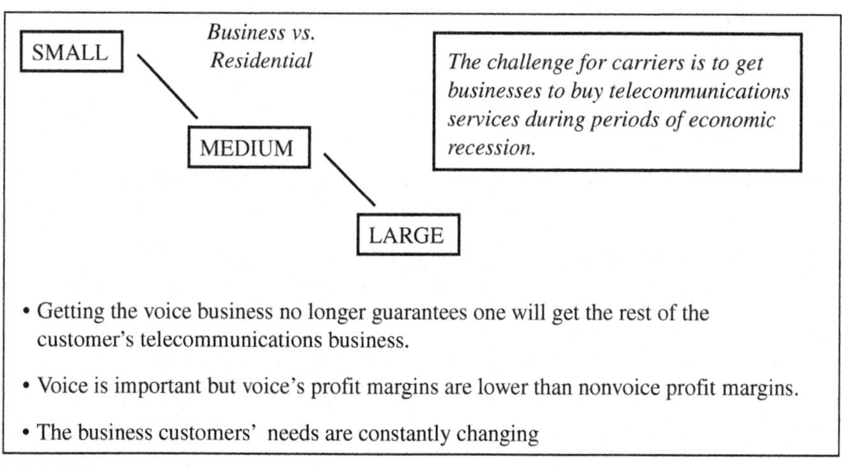

FIGURE 6.4 Business customer segmentation.

the operating environment for a carrier has changed, as has the carrier's approach to the business market. What has become apparent is that determining the importance of the customer is now based on a number of factors other than the number of lines:

- The size of the customer—number of employees
- The type of service being provided to the customer. What are the chances of upgrading service to something else?
- The quantity of the service being provided
- The profit margin of the service
- The potential for future business, also known as the upside of the account
- The type of customer—whether or not the customer is low-end or high-end. Low-end customers usually want voice-only services, not even voice mail. The high-end customers want multiple voice features and data services. High-end customers will pay more than low-end customers will.
- The location of the customer—this may have an impact on the types of additional customers a carrier will obtain. If the carrier has a small regional or local play, serving the Wall Street area, having most customers in Delaware certainly will

not help it sell services in the Wall Street area. Carriers sell services in areas where customers are located.

Getting the voice business no longer guarantees getting the rest of the customer's telecom business. Figure 6.5 is a rendering of business customer segmentation. Today, carriers tend to have two customer classifications: small and large. Based on the carrier's portfolio of services, various subcategories of voice and data can be created.

One might assume that only large nationwide carriers would be successful; in reality, there are dozens of small local and regional carriers doing very well serving the 100-line and smaller market. These small "mom-and-pop" telephone shops (CLECs) do not classify customers as small, medium, or large. The customer base of the small CLECs is not large enough to classify.

Segmenting customers in size groups is a way of focusing customer care and field operations resources. Business customers expect to have people dedicated to their accounts. Customers of certain sizes will have common needs. By creating dedicated customer people focused on meeting the needs of business customers of a certain size, the carrier will be able to nurture skill sets in a single group of customer care specialists. Moreover, network operations, switching and routing control,

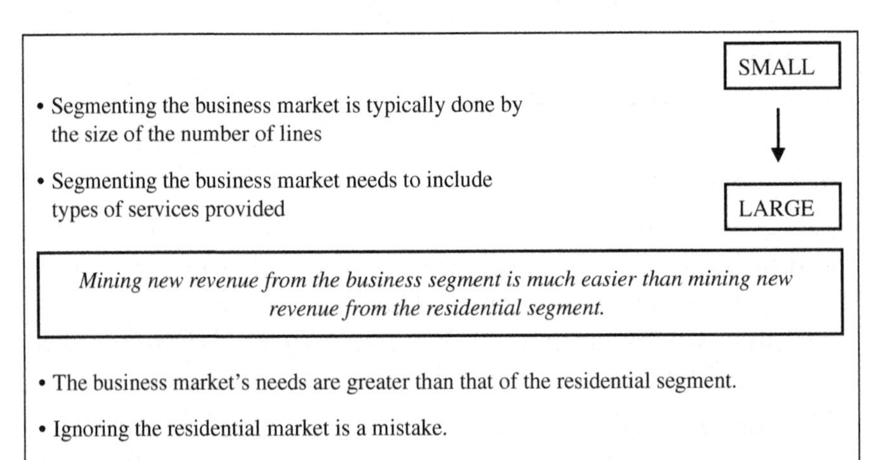

FIGURE 6.5 Business customer segmentation.

and data center management will be able to nurture and train specialists to meet the unique needs of the business customer.

As I noted, business customers of certain sizes tend to have the same kinds of telecommunications needs. Creating dedicated customer care and operations support is important to maintaining customer accounts. Knowing the customer in a way the customer perceives as dedicated and personal is more than half the job in keeping the customer happy. Segmentation of the business customer in this manner is critical in achieving the right relationship between the carrier and the customer. Business customers are no longer as loyal to carriers as they were in the early 1980s and earlier. Customer care is paramount and segmentation facilitates the customer care process.

BUSINESS SERVICES. Business customers have a variety of needs, ranging from simple voice to complex data requirements. The telecom carrier must offer a variety of voice services based on a number of basic features. These services can contain one or more features. The following is a representative example of the more popular network-based subscriber voice services.

CALL DELIVERY (CD). CD permits a mobile subscriber to receive calls to his or her directory number while roaming in a wireless network.

CALL FORWARDING—BUSY (CFB). CFB permits a called subscriber to have the network send incoming calls addressed to his directory number to another directory number (forward-to number) or to his designated voice mail box, when the subscriber is engaged in a call or service.

CALL FORWARDING—DEFAULT (CFD). CFD permits a called subscriber to send incoming calls addressed to his directory number to a designated voice mail box or to another directory number (forward-to number), when the subscriber is engaged in a call, does not respond to paging, does not answer the call within a specified period after being alerted or is otherwise inaccessible (including no paging response, the subscriber's location not known, the subscriber reported as inactive, or call delivery not active for a roaming subscriber, etc.).

CALL FORWARDING—NO ANSWER (CFNA). CFNA permits a called subscriber to have the system send incoming calls addressed to his directory number to another directory number (forward-to number) or to the designated voice mail box, when the subscriber fails to answer or is otherwise inaccessible (including no paging response, the subscriber's location not known, the subscriber is reported as inactive, or call delivery, not active for a roaming subscriber, Do Not Disturb active, etc.). CFNA does not apply when the subscriber is considered busy.

CALL FORWARDING—UNCONDITIONAL (CFU). CFU permits a called subscriber to send incoming calls addressed to his directory number to another directory number (forward-to number) or to his designated voice mail box. If this feature is active, calls are forwarded regardless of the condition of the terminating end.

CALL TRANSFER (CT). CT enables the subscriber to transfer an in-progress established call to a third party. A third party is defined as an individual not involved in the initial call. The call to be transferred may be an incoming or outgoing call.

CALL WAITING (CW). CW notifies a controlling subscriber of an incoming call while the subscriber is in a two-way call. The controlling subscriber can either answer or ignore the incoming call. If the controlling subscriber answers the second call, he may alternate between the two calls. Theoretically, call hold is an inherent capability of call waiting.

CALLING NUMBER IDENTIFICATION PRESENTATION (CNIP). CNIP provides the number identification of the calling party to the called subscriber. One or two numbers may be presented to identify the calling party.

CALLING NUMBER IDENTIFICATION RESTRICTION (CNIR). CNIR restricts presentation of the subscriber's CNI to the called party. When CNIP was first deployed on a commercial basis, CNIR was just as popular among wireline customers. This has now shifted so that CNIP is far more popular. We can only assume that people who have nothing to hide do not mind having their numbers displayed.

CALLING NAME IDENTIFICATION PRESENTATION (CNAIP). CNaIP provides the name identification of the calling party to the called subscriber. This is not a popularly deployed service at this time.

CALLING NAME IDENTIFICATION RESTRICTION (CNAIR). CNaIR restricts the presentation of the calling party's name to the called subscriber.

CONFERENCE CALLING (CC). CC allows a subscriber to have a multiconnection call, i.e., a simultaneous communication between three or more parties (conferees).

THREE-WAY CALLING (3WC). 3WC gives the subscriber the capability of adding a third party to an established two-party call, so that all three parties may communicate in a three-way call.

DO NOT DISTURB (DND). DND prevents a called subscriber from receiving calls. When this feature is active, no incoming calls will be offered. DND also blocks other alerting, such as the Call Forwarding—Unconditional abbreviated (or reminder) alerting and Message Waiting Notification alerting. DND is also a feature on some telephone devices, rather than a network-based feature.

MESSAGE WAITING NOTIFICATION (MWN). MWN informs subscribers when a voice message is available for retrieval. It may use a pip tone, an MS indication, or an alert pip tone to inform a subscriber of unretrieved voice messages. MWN has no impact on a subscriber's ability to originate or receive calls.

SELECTIVE CALL ACCEPTANCE (SCA). SCA is a call-screening service that allows a subscriber to receive incoming calls only from parties whose Calling Party Numbers (CPNs) are in an SCA screening list of specified CPNs. Calls from CPNs not on the SCA screening list and calls without a CPN are given call refusal treatment while SCA is active.

SHORT MESSAGE SERVICE—POINT-TO-POINT (SMS-PP). SMS–PP provides bearer service mechanisms for delivering a short message as a packet of data between two service users, known as Short Message Entities (SMEs). SMEs are SMS endpoints capable of composing or disposing of a short message.

One or both of the service users may be a mobile station. The data packets are transferred transparently between two service users. The network or destination application generates negative acknowledgments when it is unable to deliver the message as desired. The destination application may respond with an automatic acknowledgment and may include application-generated or user-provided information.

MESSAGING DELIVERY SERVICE (MDS). MDS permits attempts to deliver pending voice messages to a subscriber on a periodic basis until the subscriber acknowledges receipt of the messages.

PAGING MESSAGE SERVICE (PMS). PMS permits attempts to deliver paging messages to subscribers (via the SMS) on a periodic basis until the subscriber acknowledges receipt of the message.

VOICE MAIL. Voice mail provides the subscriber with services that include not only the basic voice recording functions, but also time-of-day recording, time-of-day announcements, menu-driven voice recording functions, and time-of day-routing. VM was first deployed in a ILEC switch during the early 1980s. The popularity of the answering machine limited VM's growth, but VM has seen enormous growth as a value-added service in the wireless (cellular, PCS, and paging) industry.

FAX MAIL (FXM). Fax mail provides the subscriber with services that include the ability to store faxes received and forward faxes recorded, based on a number of subscriber-set parameters.

The popular data services are all currently focused around the Internet. Before the Internet, carriers could sell a T-1 transmission facility to a business for its use as a dedicated wide area network connection. The carrier did not process any of the customer's information. Instead, it provided connectivity. Data was considered a nonvoice service. Before the Internet, there was not much a typical individual user could do with simple text data. Remember that in the 1980s, many companies still had typing pools where letters could be typed by typists using word processors. Email was not even a dream; in 1985 no one in business even knew what it was.

The Internet proved to be the kind of fundamental change the telecom industry needed to cause a change in the way carriers perceive data. As I noted in, *M-Commerce Crash Course*, the Internet became the home for telecommunications technology mavericks and it unleashed the creative genius of many technologists. An Internet-based carrier is also called an Internet Service Provider (ISP). Incumbent LECs (ILECs) have constructed networks to support high-speed and large-bandwidth services. The ILECs and many CLECs provide Internet access and a variety of Internet-based services. Types of services and capabilities provided by Internet-based carriers include:

- Email
- File storage—today called a *storage access network* (SAN)
- Web site creation
- World Wide Web (WWW) access
- Internet portal access
- VoIP (Voice over Internet Protocol)

EMAIL. Email is to the Internet what plain old voice service is to telephone and cellular businesses: the staple of the Internet. Email has assumed an identity that cannot be easily equated to that of voice in the traditional telecom business (telephone and cellular). Email is unique in its ability to allow a user to be brief and to say nothing of substance. Not even a short voice conversation over a telephone or cellular handset can boast that kind of communication brevity. Email has added a whole new dimension to communication between parties. It is both more and less intimate than letter writing.

Email literally allows one to carry on a conversation in near-real time, to engage in quick discussions about important or frivolous topics. Copies of letters and other documents can be transmitted with an email. The use of instant messaging between subscribers (customers or users) of the same ISP enables real-time communication.

FILE STORAGE. File storage refers to the storage of files that belong to ISP subscribers and is performed to a limited extent today by ISPs. Files are typically stored within email messages and not as separate archival data backups. Today the majority of files an ISP agrees to store are placed only in temporary email storage areas under the user's email logon name.

As I noted in *M-Commerce Crash Course*, some ISPs offer archival data storage, a service that a corporate management information system (MIS) organization may offer to its corporate users. The new marketing term for this service is the storage access network. Boil this down to fundamentals and we still have a data storage company.

WEB SITE CREATION. Web site creation is a service that many ISPs offer to subscribers. A subscriber pays a small fee to use software tools that can be found online with the ISP. These tools enable a customer to create a personal Web site with pictures, font styles, colors, and interactive icons. This is an interesting, useful, and helpful tool that allows the ISP's customers to take advantage of the Internet by enabling them to become more than just simple email customers. It enables the customers to become Web information providers themselves.

Web site creation promotes the use of the Internet. Even better, for a small fee, the ISP becomes the host of the customer's Web site. The ISP's computers become the physical address of the customer's Web site. Customers need not purchase or manage their own servers. The Web site creation tool generates Internet usage. Web site creation is no longer a hot marketing item, but in the telecom business, many services have a way of resurfacing with a new "coat of paint."

WORLD WIDE WEB (WWW) ACCESS. The World Wide Web is not a single object, but a collection of Internet sites. The ISP enables a user to access multiple Internet sites. Imagine you are in orbit about the planet Earth. You look down and see thousands of Internet host computers scattered all over the planet. Figure 6.6 is a depiction of the various host computers scattered across the Earth. If you were to connect all of the host computers using a pencil, the picture would look like a spider web; hence the use of the word "web" as it relates to the Internet. The term World Wide Web is an accurate description of the relationship between all the Internet sites. Figure 6.7 is a representation of the web I have described.

The Internet has become so pervasive that countries in which there is little if any wireline or wireless teelcommunications infrastructure will instead have widespread Internet access.

FIGURE 6.6 Internet sites all over the planet.

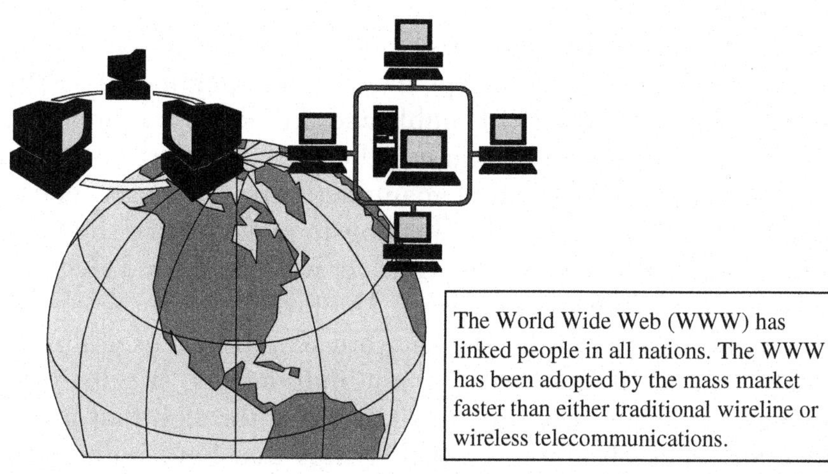

FIGURE 6.7 The World Wide Web.

The ability of carriers to enable users to communicate with other ISPs is a product of the ISPs. Access is a product. Users have no need to purchase service from an ISP that cannot access the site of a company selling products they want. Although an ISP may not be selling any products beyond email, an ISP that simply offered email and not the ability for its users to surf other Internet sites would not be in business long. In the beginning, ISPs provided only email and limited Web site access unless they hosted the Web sites for various companies. There was limited access to other sites during the early days of the Internet because of the limited penetration of computers and limited existence of commerce-related Internet sites.

Telecommunications networks, including data networks like the Internet, form webs. Today the Internet is attempting to create a level of traffic management order within itself, a "hub" concept, where single ISPs work to act as aggregators of information for users. This concept, heavily used by other telecommunication carriers (e.g., wireline and wireless carriers), has taken on the form of the *portal* in the Internet world, where a hub acts as a focal point of traffic. There will be more on the portal later in this chapter.

Web access enables users to surf the ocean of information that exists at the various Internet sites. Companies, libraries, schools, universities, and other institutions that have their own Internet sites provide this information.

INTERNET PORTAL ACCESS. As I noted in *M-Commerce Crash Course*, the portal is a concept that became popular in 1999. Early in that year, the IEEE published articles in a number of their society magazines detailing the research performed by various computer equipment manufacturers. The issue being addressed in these magazines was how to make surfing the Web more efficient for the user. The answer was simple and elegant: build Internet Web site gateways. The problem with nearly all the search engines is that the keyword search by a user almost always results in hundreds, if not millions, of site listings. People do not have the time to sit and surf through that many sites. Searches often end up listing sites that have nothing to do with the topic of interest.

The portal is an organized way of facilitating user searches: controlled and managed information access. Portals are designed to focus on specific needs. For example, a children's portal may focus on new toys being marketed this year. Another portal may focus on sports cars made in the United States between the years 1999 and 2002. Another portal may focus on charities in a given geographic area. Another portal may focus on specific medical issues. Portals can to support a variety of products, issues, and activities. Figure 6.8 illustrates this point.

Portals are managed so that one can visualize the portal-to-Web site relationship as an organization chart; specific positions are related to other positions (just follow the lines of organization). Another way of viewing a portal is as a highway toll both. As multiple lanes of traffic approach toll booths, the cars are often forced to merge into fewer lanes. Portals are points of search and information access aggregation. Figure 6.9 is an illustration of the portal concept.

The portal is an excellent application of basic traffic management, but it has downsides for the user. The portal is managed access, therefore the user must trust that the portal

Portals are points of search and information access aggregation.

- Internet portals are gateways to a universe of Web sites.
- Portals can be likened to the hub of a wheel with spokes—the spokes are highways to Web sites.

FIGURE 6.8 Internet portals.

The portal is like a traffic light.

The portal is an Internet application that has attempted to bring organization to the surfing the Web.

The portal can serve specific commercial or nonprofit interests.

Portals can be managed by any company for any cause.

Example:

ADoorAjar—not-for-profit portal championing literacy

FIGURE 6.9 The portal.

manager has established links to the appropriate sites; otherwise, the user will be forced to search through hundreds of sites. This is a challenge for the portal manager.

VOICE OVER IP (VOICE OVER INTERNET PROTOCOL). Voice service has the broadest appeal to the marketplace. It is the lowest common denominator in the telecom service provisioning equation, because it is what all customers looking for. Internet companies/carriers are all seeking to look like the telephone company by providing voice. Remember that not everyone is using the Internet. The power of the wireline telephone company is its familiarity for all people. The telephone company's primary service is voice. To the nontechnical user (the majority of the marketplace), voice is the big service. ISPs are aggressively seeking to sell voice over the Internet in order to broaden the Internet's appeal and to generate more revenue.

GOVERNMENT

In the retail telecom segment, government is a big customer. It is defined as any state, local, provincial, regional, or national governmental agency or bureau. The government is a big user of:

- *Voice.* Citizens always call the government and vice versa.

- *Data.* Dedicated transmission of data between data centers
- *Internet.* Citizens surf agencies' Web sites.

The government must be accessible to citizens, therefore telecommunications are essential. Telecom contracts with government agencies tend to be large awards. More often than not, multiple agencies will award telecom contracts to the same carrier. Government telecom contract awards are not simple to provide. Stringent performance parameters must be met. The telecom carrier providing service to a government agency will find the government a tough customer to please and one not tolerant of poor service and delays in repairs. This is the one customer for whom vendor and carrier will work together to ensure the carrier wins the contract.

The U.S. federal government is considered by all U.S. carriers the nation's single largest telecom user. In the early 1980s, the Federal government instituted a program called the Federal Telecommunications System 2000 (FTS2000) to bring digital voice, video, data, and long distance transmission services to all of its agencies. During that time, local and long distance worked together to provide these services. The long distance companies did not and still do not have the penetration into enough local markets to provide all such services. By working with local telephone companies via contract, long distance companies can meet the needs of the Federal government. FTS2000 (today called FTS) is akin to a large private network, and meeting the requirements of FTS2000 is not a small or easy task. Figure 6.10 is an illustration of FTS.

WHOLESALE TELECOMMUNICATIONS MARKETPLACE

The wholesale telecom marketplace refers to the provisioning of services to the user via a third party. The company that purchases the services in order to sell to a user is called the reseller. The company that sells the services to the reseller is called the wholesaler.

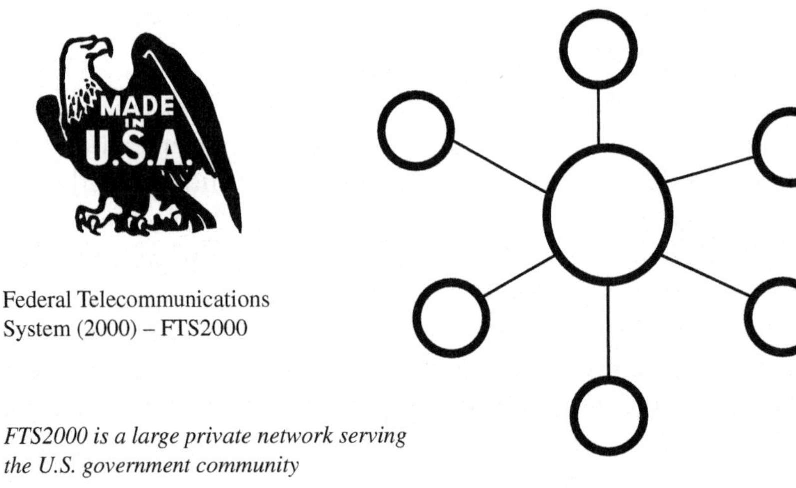

FIGURE 6.10 FTS.

The wholesale market for the telecom industry can be divided into vendors and carriers. Typically, descriptions of the marketplace refer to the carrier community, though manufacturers of hardware and software also have their own wholesale segment. Vendors use agent companies to sell their equipment; these agents may be carriers or other vendors. The OEM vendor is essentially a reseller of the small manufacturer that is seeking a way to distribute its product. This section will focus on wholesale services. Figure 6.11 is an illustration of the wholesale concept.

The concept of wholesale services has been around since the 1980s. As I noted in *Telecommunications Internetworking,* there are ILECs that provide various kinds of support to other carriers. I have always preferred the term *infrastructure support,* although the more popular term is wholesale services. *Telecommunications Internetworking* describes the mechanics of how to provision wholesale telecom services.

The role a carrier plays within the larger network of networks has a great deal to do with how it is interconnected with other networks. How two carriers are interconnected will determine the kinds of telecom services one carrier will offer the other. Figure 6.12 illustrates how carriers work with one another in a wholesale relationship.

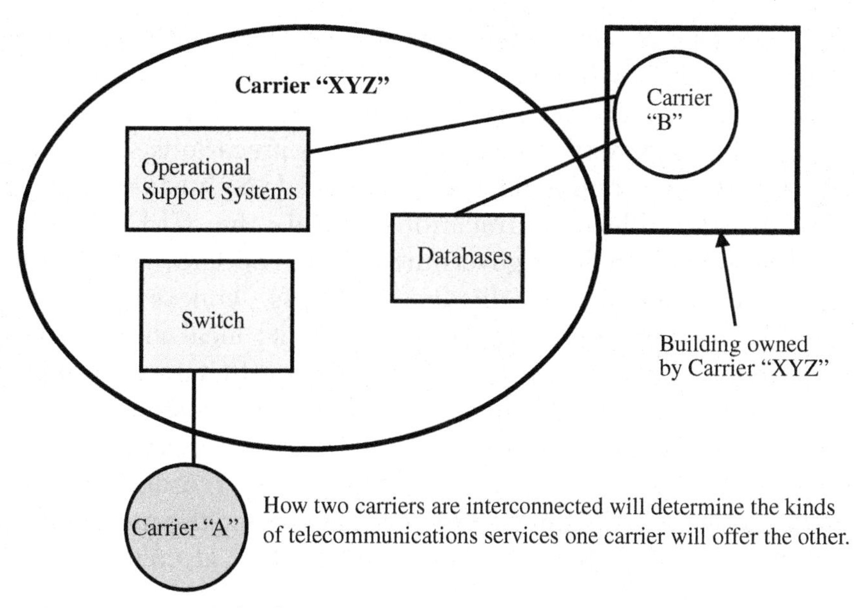

FIGURE 6.11 Wholesale concept.

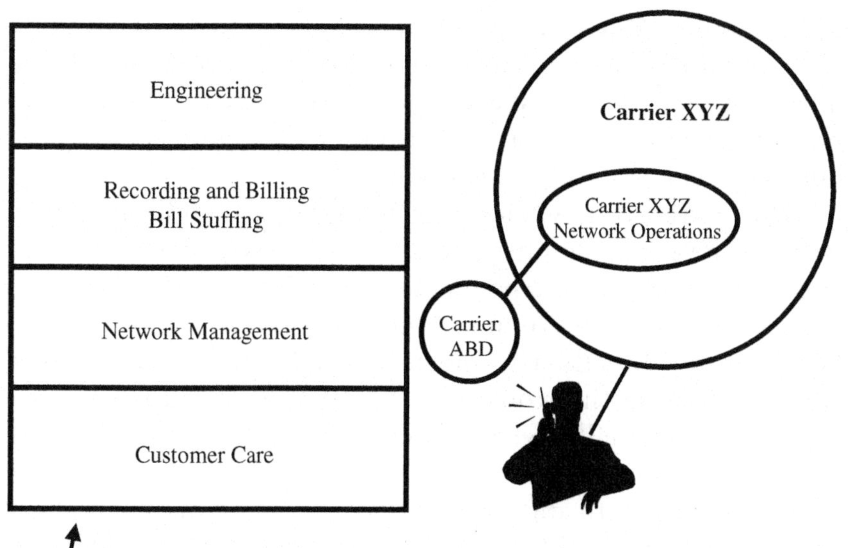

FIGURE 6.12 Carrier wholesale relationships.

Nonfacilities-based and facilities-based resellers are in a wholesale relationship with the ILECs. This reseller is called a Competitive Local Exchange Carrier (CLEC). The carrier providing wholesale services finds itself in a precarious situation. The CLEC environment is not a typical U.S. business situation. In the telecommunications world, the CLEC exists because of U.S. federal government law. The U.S. ILECs are seeking to enter the long distance business. Unless the ILEC can demonstrate there is competition in its local market, the U.S. Federal Communications Commission (FCC) will prohibit it from entering the long distance business. The Telecommunications Act of 1996 requires carriers to meet the criteria of a 14-point checklist, state by state, in order to enter the long distance business. One criterion is local competition. The law requires that there be significant local competition, fostering the existence of CLECs.

The CLEC is typically a reseller of ILEC services. There are CLECs that build networks, provide dial tone, and various services to the customer over their own transmission facilities. However, the majority of the CLECs use the ILECs' transmission networks and switches to provide services. The law requires ILECs to discount services substantially to CLECs. The 2001 spate of CLEC failures may seem to indicate that the wholesale telecommunications business is a failure, but these failures are not the result of a lack of market for CLECs but rather of poor business plans and poor execution.

The market for wholesale services is a complicated one, with low revenue margins. The telecom wholesale business was originally engaged in dependable transmission facility leasing. Telecom carriers believe in staying close to customers, controlling accounts directly. This means a reseller eventually needs to move away from the wholesaler. CLECs using ILEC switches or routers eventually need to provide their own switching and routing equipment.

CLECs need to differentiate themselves from their competitors, but this is difficult if they need ILECs to provide services. One might be able to make a financing decision to continue using an ILEC switch or router, but the decision to have one's own switch or router will be based primarily on competitive needs.

Considering the way the telecom industry is changing in response to technological and regulatory changes, it is possible that the wholesale market will experience enormous growth.

TYPES OF WHOLESALE SERVICES

My view of the wholesale business includes all assets of the wholesaler. The equipment and processes that can be sold on the wholesale market include:

- Switches and switching
- Routers and routing
- Servers
- Antennas
- Towers and rooftop
- Databases and database management
- Operations support network services—operations support is normally defined as methods, procedures, and systems that directly support the operations of a telecom service provider.
- Network management
- Recording and billing (also known as call detail recording systems)
- Colocation—a regulatory construct in which an ILEC provides space for the CLEC's equipment or frame appearances. The following items can be provided under collocation.
  - Electric power
  - Water
  - Heating ventilation and air conditioning
  - Building space/floor space

Far more detail on this subject can be found in *Telecommunications Internetworking*. However, I should add that a telecom service provider is an information engine capable of performing a number of activities on behalf of other carriers. These activities can include:

- Bill stuffing
- Data warehousing
- Customer care support
- Maintenance and repair
- Installation

Whether or not a carrier will provide these services to another carrier is a matter of business policy for that company. Figure 6.13 is an illustration of the kinds of wholesale services a carrier may offer.

MARKET FOCUS

Carriers must avoid using technology that outpaces the needs of their customers. Carriers have been known to become so enthralled with new technology that they forget to focus on the customer. The danger of not staying aligned with the carrier's sales and marketing force is of creating services that no cus-

In order for the carrier or the vendor to stay competitive, it must stay slightly ahead of market expectations.

Wholesale has expanded to also mean real estate and utilities.

- Electricity
- Building Space
- Water
- Rooftops
- Switching
- Billing

FIGURE 6.13 Carrier wholesale services.

tomer wants. This should not be construed as saying that the carrier should not be one step ahead of market expectations. In order for a carrier or vendor to stay competitive, it *must* stay slightly ahead of market expectations.

The goal for a carrier is to stay market focused. Meeting and synchronizing the needs of business and the residential community is a major challenge for carriers, because the two customer communities have different interests—which is why most CLECs focus on just one customer community. Many carriers will focus not on the market but on the sales figures. What I am addressing is a carrier's desire and need to meet its financial commitment to its shareholders. Meeting the dollar commitment is always a priority, but equally important is ensuring that the business is providing the kind of service(s) that will satisfy customers' needs today and in the future. Many carriers focus on the short-term sales commitment and not on the future need to continue providing attractive services.

Customers do not care if a carrier is profitable; they care about having their needs met. A carrier (also a customer) does not care if a vendor is meeting its revenue and margin commitments, because all a carrier wants is equipment and software that meets its needs. Carrier and vendor executives must always remain aware of their carrier's financial status, but financials cannot be the only factor guiding company decisions. Other things to be considered are sales, marketing, finance, product management, customer care, etc. Meeting the needs of the customer is a balancing act of meeting the needs of company shareholders and of those who purchase the company's services or products. A telecom executive must be focused on all aspects of the company's operations, which ultimately translates into meeting the needs of the customer. Everyone to whom the company is responsible for reporting or responding is in some sense a customer of the company.

In order for a carrier to stay market focused, it must first identify the market it serves. Once the market is defined, staying on target is an ongoing task of self-examination. Straying from market focus is easy. Companies often become self-absorbed in their own accomplishments and forget to look at their customers' changing needs. Defining the market is not simply a matter of carving up a town or city by using demographics as guidelines.

Defining a market is an exercise in observation. I noted that the telecom marketplace is divided principally into retail and wholesale areas, but these are just market areas. An industry can be a market. A city can be a market. A group of people in a certain age group can be a market. A group of people in a certain profession can be a market. The parameters by which one can define a market are numerous. The boundaries of a market can vary widely. A telecommunications carrier needs to divide the market so that all segments can be served adequately.

The need to maintain a market focus is a struggle between sales objectives and meeting customers' needs. The difference between a company whose sole focus is sales and one that focuses on the customer is the difference between one that is in business and one that is not. Wireless carriers lose millions of customers despite low prices because their service is worse than that of other wireless carriers. Wireline telephone companies maintain huge customer care organizations because customer contact and the perception customers have about how their troubles' are being addressed are paramount. Figure 6.14 is an illustration of the balance between sales and marketing.

MARKET SEGMENTATION

In general, the telecom marketplace is an alluring one—because high technology has a way of attracting interest—but every area of the marketplace is not equally attractive. In order to increase chances for a larger return on investment, we segment the markets into subcategories, in which the chances for profit are greater. In the world of telecom service provisioning, market segmentation is an ongoing exercise in one's sharpening understanding of customers.

Segmenting the market requires creating strategies that support different products for different segments. Call Forwarding and Call Waiting are two different services designed to perform two different functions. Call Forwarding permits a called subscriber to send incoming calls addressed

FIGURE 6.14 Sales focus versus market focus.

to his directory number to another directory number (forward-to number) or to his designated voicemail box. If this feature is active, calls are forwarded regardless of the condition of the terminating end. Before voice mail became so common, wireline telephone companies found Call Forwarding very popular among business people (and also among the criminal element working to have critical telephone calls diverted to other locations).

Call Waiting provides notification to a controlling subscriber of an incoming call while the subscriber is on a two-way call. Subsequently, the controlling subscriber can either answer or ignore the incoming call. If the controlling subscriber answers the second call, he may alternate between the two calls. Call Waiting is popular among residential customers who cannot afford two voice lines in the house. It enables a single line to serve homes where there are multiple telephone users. It enables users to ask the second incoming calling party to wait or call back. In other words, Call Waiting enables a residential user to answer a call with a real person rather than an answering machine. To many callers, a response is better than leaving a message on an answering machine. How does the carrier segment this customer base?

To segment the customer base, the carrier must understand how the various customer groups use telecommunications in their activities, their daily usage, device preference, and their called or communicated parties. The same is true of the way vendors sell services, software, and hardware to carriers. Vendors group their carrier customers into categories based on the types of services the carrier sells, or even on the size of the carrier's subscriber base. Segmenting may result in carriers that sell services to the mass market or to a niche market.

The process of segmenting the telecom market is part science and part art. The science part of segmentation is described in the next section. The art is experience, and must be gained. Figure 6.15 is an illustration of how segmentation works in the carrier environment.

STEPS IN SEGMENTATION

The process of segmentation can be divided into four distinct steps, described below and in Figure 6.16:

1. Identify potential segments
2. Select segment groups

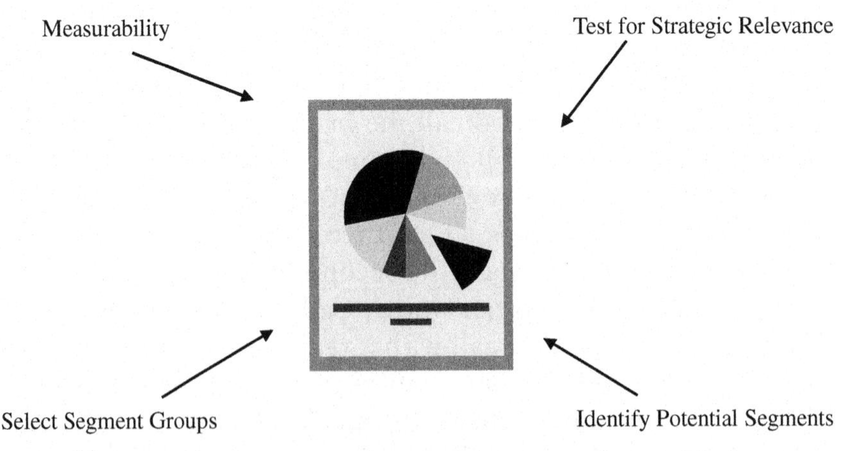

FIGURE 6.15 Segmentation in telecommunications service provisioning.

Segmentation requires the carrier and vendor to ask a series of questions that enable them to identify the customer segment and apply some kind of test to their business-customer concepts.

The carrier and vendor need some way of testing their views otherwise money may be lost on an idea that looked promising at first but proved to be a bad idea.

- Identify Potential Segments
- Select Segment Groups
- Test for Strategic Relevance
- Measurability

FIGURE 6.16 Steps in segmentation.

3. Test for strategic relevance
4. Measurability

IDENTIFY POTENTIAL SEGMENTS

This sounds easier than it really is. Most marketers use demographics as a way of dividing the market into appropriate customer segments. However, selling something as complicated as telecom services, location-based services, Internet-based services, satellite news, and paging services requires more than just typical demographic segmenting. A vendor may wish to segment the population of customers into the following categories:

- Carriers with CTOs who have worked with each other
- Carriers with research staffs
- Carriers without research staffs
- Carriers using a particular network technology
- Carriers using digital technology
- Carriers seeking to use ATM backbone signaling
- Carriers that provide both local and long distance services
- Carriers that serve a particular segment type

This list could be even longer. Depending on the vendor's creativity, the "pie" can be divided into many slices. In fact, a creative vendor should be able to "slice and dice" the market in a number of unique ways.

The most important question and usually the one everyone forgets to ask. Carriers and vendors have been known to rush ahead with advertising campaigns selling various services and products. What inevitably occurs is that they realize they are selling to a specific customer segment with no desire to purchase. Though they recognize this problem was never adequately approached, they nevertheless continue barreling ahead with sales and advertising campaigns. One of my favorite stories is about the first time the wireline telephone industry started offering Caller ID. The concept was new and hundreds of minds were working at finding practical applications for this service in everyday life. The results of this search were being used in ad campaigns. Security from crank callers was the big application and the service was being marketed to everyone regardless of customer type.

One day someone got the bright idea that the telephone companies could offer, for a fee, Caller ID Block (Restriction) to those who do not want their telephone numbers known when they make a telephone call. It may seem bizarre for a telephone company to sell to those protecting their privacy from strange callers and to those who want to make telephone calls but not have their true identity known. It was considered odd even in the 1980s, but customers eventually fell in love with the services. The penetration is still not overwhelming, but it is growing. One telephone company automatically provided Call Waiting to all customers for a fee, unless users requested removal from the provisioning list. It gave users the ability to suspend Cancel Call Waiting for very brief periods of time for a fee. That eventually changed to providing Call Waiting to customers as an optional service and allowing those same customers to suspend Call Waiting for brief periods of time for the same single fee. The way Call Waiting and Cancel Call Waiting had originally been provisioned was comical. No one took the time to understand who could best benefit form Call Waiting. The assumption was that all customers used telecom services in the same way. The wireline telephone companies could have increased their profits if more time had been taken to understand the market.

In the past, very few people in telecommunications took the time to understand how different segments of the customer base purchased wireline telephone services. Many of the services we use today were new in the 1980s, and every customer seemed interested in purchasing these services, regardless of the situations in which the services would be used. The post-divestiture AT&T arena (after 1984) was a brave new world for customer, carrier, and vendor alike. One assumption was that all market segments purchased services the same way. Another assumption was that customers would buy anything they were offered. Both assumptions were erroneous, but understandable, because there was no history to draw from.

Ultimately, the most important issue is to identify customer, before any other substantive sales, marketing, and advertising work is done. Answering this question is a major problem for most companies.

Identifying segment groups can be done by identifying different customer groups and the responses of those groups to the service or product, and then finally identifying markets. A list of the variable parameters used to identify segments includes:

- Consumer markets identifiers
  - Demographics
  - Lifestyle
  - Profession
  - Culture
  - Language
  - Entertainment habits
- Industry market identifiers
  - Industry type

  - Purchasing patterns
  - Suppliers
  - Service types
- Consumer behavior
  - Heavy users
  - Light users
  - Reactions to changes in the economy or service
  - Personal or business use

The three principal attributes used to describe a customer segment are: consumer markets, consumer behavior, and industry markets. These attributes also can be applied to carriers by vendors. Figure 6.17 illustrates the steps in identifying segments.

At the end of the analysis, the company will have a series of profiles grouped according to common characteristics. Customer profiles contain relevant information about various customer segments. The characteristics used to define the segment groups will vary depending on the company's needs. It is important to decide how one selects traits that will be used to create groups of segments. Note that the method of selecting common traits can lead to error. The next step is to use the customer profiles.

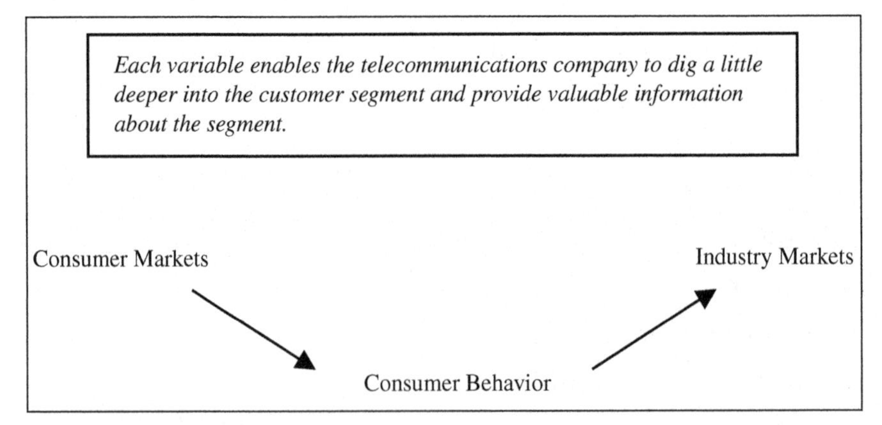

FIGURE 6.17 Identifying segments.

SELECT SEGMENTS

The goal of selecting customer segments is to associate attributes with consumer groups. The object of this step is to take the profiles and burrow deeper into the analysis of the customer segments. The carrier or vendor will establish a set of parameters by which it will use the customer profiles it has created. The analysis can be described as "WIFM," pronounced "whiff–fim" and standing for "What is In it For Me?". The term is an old sales adage used by experienced salespeople to check their sales and tactical strategies. Those who work to earn a living guide their actions by their own personal agendas. Consumers buy for a reason which certainly is not to help the salesperson meet a monthly quota. To deny this would be folly; look at what motivates you in your everyday interactions. "WIFM" can be described in very discrete parameters that are variable in nature:

- Application
- Benefits gained or sought
- Market and environment sensitivity
- Costs

These variables describe how consumers make decisions; they are used to establish the segments in which to pursue business. Selling to a user means understanding what motivates him. With this focus on the customer, the following describes "WIFM" and how it is used to create customer segments.

Application refers to the use to which a particular telecom service is put. Services and products only have value if there is a use for them. The application may be something as simple as a tiny handbag that can be carried over the shoulder and as a backpack. This discussion of application leads directly into benefit.

Benefits gained or sought refers to benefits derived from using a product. Different types of consumers will be able to gain the same benefit even though the product and service is being used to support different business types. Carriers group customers into categories of users who gain a certain benefit from a service. For example, voice mail service is used by business and residential customers, but primarily in business. Residential users of voice mail are usually high-end business professionals who employ the service to support their business needs at home as well as at the office. Business users of voice mail are small businesses who cannot afford their own telephone systems; these may be Small Office Home Office (SOHO) users as well as small business offices. The benefit is not just the ability to store voice messages but also to reduce cost and optimize time management. These small businesses are likely to be service companies.

The carrier groups are segments of small businesses with a need to stay in contact with customers constantly but not necessarily in real time. A car service would not find this service useful because customers want instant communication. A travel office or a small engineering consulting firm would find voice mail useful. Services and products must be useful if they are to be purchased by consumers: usefulness is key to the success of a product on service. However, usefulness does not guarantee a carrier's continued success if the service can be easily replaced by a machine like an home or office answering machine. The consumer will shift from one service to another or a replacement product if the price changes up or down. This leads us into the next variable, market and environment sensitivity.

Market and environment sensitivity refers to the effect of changes in price, appearance, quality, loyalty, brand perception, and performance on a consumer's purchase of the service or product. Every consumer of telecom services is price sensitive. In an economic environment where cost is a major concern, a shift in price of 1 percent can be the difference between a customer meeting or not meeting budgetary goals. Other parameters like quality, loyalty, appearance, brand, perception, and performance, also play a role. Carriers and vendors have found that this particular variable—price sensitivity—has an overall effect on consumer's behavior.

Consumer loyalty is another major factor. After the 1984 divestiture of AT&T, millions of consumers refused to consider purchasing service from MCI simply because of loyalty to AT&T. Nearly 20 years later, there are tens of millions who now purchase from MCI, Sprint, and other carriers besides AT&T; however, there are also tens of millions of people in the U.S. who believe that AT&T still operates all the local telephone companies. This may be more than just loyalty: it is a matter of national memory. In the United States, there is not an adult alive who cannot remember a time when there not an AT&T. From a marketing perspective, this could be called brand identity. These two factors have both a measurable (price-related) and perceived (brand and loyalty-related) nature.

Costs refer to costs to the carrier, vendor, and consumer. Carriers and vendors work to provide services and products in a fashion that does not require spending lots of money fine tuning a product or service for customers. When you provide a service or product to different customer segments, there are differences in requirements. These lead to cost differences in providing the service or product. Salespeople will initially target "big low-hanging fruit" customers because they not only need to make a quick and large sale, but also because the cost to sell to a small customer versus a large customer tends to be the same. If you are going to spend money, spend it on the big but easy sales target. In the case of the consumer, the cost of a telecom service is a part of the budget or operating costs. Consumers of a certain size and industry type may have similar operating cost objectives and can be grouped in the same categories despite differences in industries.

As I noted, when you take all the variables described, you create a profile of the customer segment. This information will be used to create sales and revenue projections and justify the company's business strategy. It ultimately serves as a critical element for the business strategy of the company. Figure 6.18 is an illustration of how segments are selected.

How Consumers Make Decisions

- Applications
- Benefits Gained or Sought by the Company
- Market and Environment Sensitivity
- Costs

- Carriers and Vendors need to understand how consumers make decisions
- Carriers and Vendors need to sell a product or service that is useful and meets the consumers' other needs.

When selecting the segments, the carriers and vendors gather information about the target segments and then make their selection based on the segment's revenue potential and growth over a period of time.

FIGURE 6.18 Selecting segments.

TEST FOR STRATEGIC RELEVANCE

The variables—application, benefits gained or sought, market and environment sensitivity, and costs—together create a picture or profile of the customer. Different attributes of the customer are emphasized by the four variables. How does a carrier or vendor know that the picture created is accurate enough to be a useful approach to the market? This picture or profile must be tested against that which consumers find relevant to their needs.

Testing the carrier's or vendor's segment models can only be done by executing the sales process. If the process of execution is done incorrectly, a telecom company can lose its entire business. In the telecom business, companies tend to get only one chance to make a mistake so companies strive never to make a big mistake. Carriers and vendors seek knowledgeable and high-powered sales and marketing people to run their sales and marketing organizations, because experience translates into more wins than losses.

MEASURABILITY

Measurability is the last step in segmentation. During the sales process, the CFO and VP of Sales in a telecom vendor should be calculating the profit margins on each contract or on all the contracts. The only true measure of whether or not one has segmented the market properly is the final deal or set of deals (contracts). Sometimes the first contract has to be signed at a very low profit margin in order to close bigger deals in the future. The first contract is a sales tool for the next set of contracts. The only true measure of testing segmentation is success.

This may sound unfeeling, but the only thing that matters is the margin on the deal. Imagine going through a detailed analysis of how the market is being segmented. Every member of the senior staff is involved in the discussion, providing input based on years of experience. When the discussion is over, the head of sales is told to create a sales strategy. The only way the VP of Sales can determine if the sales strategy, which used segmentation as its input, works is by going through the entire sales process and closing the deal. Experienced salespeople should and will be able to determine if the manner in which the customer base was segmented was optimal long before negotiations with the customer ever begin. Imagine the head of sales telling the CEO (after the contract is signed) the company was wrong about its view of the market and could have signed a more lucrative deal if the market segmentation had been different. The head of sales should have seen this coming long before the contract was signed. Bad contracts are signed all the time, but may be avoided if the company is vigilant.

Carriers face a challenge because ultimately they copy each other. Service differentiation is a fleeting thing in the carrier community. In nearly every instance, carriers are forced to enter price wars with competitors. This may change one day because technology is changing so rapidly that it can give a carrier an edge over other carriers. However, this does not eliminate the importance of segmentation. A carrier may spend $1 billion deploying a great new Internet-based service only to discover it was wrong by one year about the marketplace's readiness. One year may not sound like a long time but one year can mean the difference between profitability and bankruptcy.

Measuring a telecom service's success requires patience. Profiting from a telecom service requires a carrier to be patient. Customer education is always a component of the telecom service business. It takes time, and that time is a cost the carrier must bear until the service becomes profitable. Measurability has a time component. Knowing when to declare a service a success or failure is based on the carrier's business objectives.

MARKET DRIVEN VERSUS SALES DRIVEN

A market-driven organization is focused on the customer. A sales-driven organization is focused on meeting "numbers." A successful company is focused on the customer. Many marketing management gurus have espoused a customer-focused philosophy. One does not need to be a rocket scientist to understand something that the corner grocery store owner has understood for hundreds of years: "You want to make money. Who has the money? The customer has the money."

There is an old sales adage that the customer is always right. I have encountered a large number of salespeople who will work too hard to show a potential customer that he is wrong and will show the customer how wrong he is about their product. At some point, the salesperson has to walk away. Annoying a customer, even a potential customer, will guarantee absolutely no business in the future. As a former colleague of mine once said, "Think about what you are going to say, because they have the money and we don't. Focusing too much on meeting your numbers and not meeting the customer's needs will guarantee that you will stop listening to the customer and eventually blow the deal. Blow the deal and we can kiss the company goodbye." You cannot make it any clearer than that.

SUMMARY

The telecommunications market is a complex environment to comprehend. Companies are both customers and suppliers. Customers of service can be companies or individuals. The business is over 125 years old and is embedded within the consciousness of the public. The telecom service business is perceived to operate in both commercial and public-rights domains. It is both technology focused and customer-care focused. Market windows are small. Timing is everything. The telecom market contains individuals, families, small businesses, large business, and government institutions. All customers are important. From a public-rights perspective, one individual customer who has suffered as a result of a service disaster can turn an entire telecom carrier on its ear.

Segmenting the market is critical to ensure providing the right set of services at the right time. Technology is changing so rapidly that understanding the market is of paramount importance. When the telecom industry consisted of monopolies, making a marketing error could easily be compensated for because there were no competitors; the market was guaranteed. Today, the industry is comprised of many carriers, and marketing and sales errors need to be minimized, if not eliminated altogether. The next step for us to understand is how we should go about organizing all this information to create a strategy. The next chapter will examine how to develop a market-driven strategy.

CHAPTER SEVEN

DEVELOPING MARKET DRIVEN STRATEGIES—HOW?

As I have noted, from the late 1990s through 2000, the telecom industry underwent an unprecedented period of enormous growth. Unfortunately, the industry was more focused on spending money and building big networks than on building what the market could use. This chapter takes a closer look at the processes involved in developing market strategies. The first thing we need to do is to create a framework by which we can begin the process of organizing a company. After we create this framework we need to create the marketing strategy.

In a Management "101" course the framework process is called *adaptive planning*. In the 1990s, the rapdily changing nature of the telecommunications business made adaptive planning a necessity. It is a framework for discussion involving multiple sources of information. The process enables multiple parties, specifically senior managers, to discuss and integrate diverse views. A structured framework is needed for such management discussions for the following reasons:

- Senior managers or other employees have opinions. These opinions need to be captured to better run the business. This framework provides an organized way of hearing the various opinions.

- Companies will be able to integrate the recent experiences of their managers. Theories and concepts are helpful but nothing is as good as real-world knowledge.
- Managers who have participated in the strategy discussion will have a better understanding of the basis upon which the decisions were made.
- Managers who have participated in the strategy discussion will feel a part of the decision-making process and therefore committed to carrying out the final decisions.

The need to integrate the management team into the strategy discussion does not remove senior management's responsibility to make the final decision. However, without the full explicit and implicit support of managers, the company runs the risk of only getting 99 percent of the management team's knowledge, savvy, cleverness, and dedication; the CEO and the board need 110 percent of the management team's brainpower.

As I noted in Chapter 5, the elements of a business strategy are:

- Market segment and size
- Value—the customer pitch or "story" that describes and differentiates the carrier or vendor from others
- Distribution—how to get the company story told; how to deliver product
- Scope and scale of execution

A business strategy is essentially a plan to achieve an objective. The process, or framework, we need to work within to create the strategy uses a planning methodology called adaptive planning. There are other tools or approaches, but the adaptive planning method works best in rapidly changing environments.

ADAPTIVE PLANNING—ORGANIZING MANAGEMENT THOUGHT PROCESSES

An adaptive planning brainstorming session of senior and mid-level managers can be likened to controlled chaos. A group of driven, intense, smart, clever, managers, who are thrusted together in a room will usually produce a fireworks display. Adaptive planning became a necessity in the telecom industry because of the dynamic nature of the business and the need to have equally dynamic managers. This planning process can be broken down into the fundamental components required to analyze any problem: assess, analyze/think, plan/decide, and implement. The adaptive planning framework is comprised of four basic components. These components are illustrated in Figure 7.1.

- Assessment
- Strategic analysis
- Decision making
- Implementation

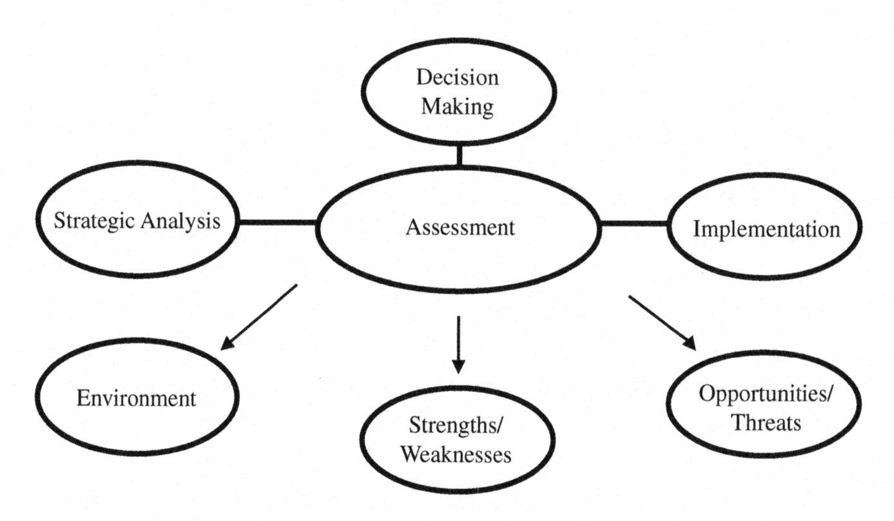

The key thing to understand in a business battle is to know your own limitations as well as your competitors' limitations.

FIGURE 7.1 Adaptive planning framework.

ASSESSMENT

Step one in the process is assessment. Assessment is both a backward and forward looking activity. It refers to an assessment of the company's strengths and weaknesses and its competitor's strengths and weaknesses. Included in this assessment step is a situational examination of the external environment. The external environment view is an examination of the competitive environment. In addition to the internal and external views, the company needs to examine the problem from the perspective of testing assumptions.

Another component of assessment is comparing the company's past performance with its success or failure in meeting its past objectives and then comparing this against the company's anticipated future performance. Before the company charges forth with its plan, it needs to think through—from a high-level perspective—the overall scope of the challenge and the environment it faces.

The key things to understand in a business battle are the limitations of the company and its competitors. Assessment can be broken down into three fundamental areas:

- Environment
- Strengths and weaknesses
- Opportunities and threats

ENVIRONMENT. Environment includes a group of factors that go beyond simple market forces; these include consumer reactions, consumer patterns, competitive action, and cultural, technological, and political issues. The economy of a country is affected by people's actions, cultural responses, and political environment. All the environmental factors described are blended together at any given time in unique ways.

In addition to the external environmental factors, a company must manage issues with its employees, corporate culture, and financial health. Environmental factors are:

- *Political.* Including regulation and legislation.

- *Economic*. Relating to the overall health of the national or local economy. If the national or local economy is bad, carriers and vendors will likely suffer because they cannot make sales.
- *Social*. Relating to the country's cultural values and beliefs. These have an impact on everything from how people eat to how they conduct business.
- *Demographic*. This can be linked to the social factor; however, it is distinct enough to warrant treatment as a category by itself. Telecom usage and purchasing is affected by a person's income, age, education level, and profession.
- *Technological*. The telecommunications business is a technology oriented business. Vendors and carriers are constantly changing the technology being sold and being deployed to serve the customer.

Figure 7.2 is an illustration of environmental factors.

STRENGTHS AND WEAKNESSES. Strengths and weaknesses typically refer to a company's self-assessment. However, I take an expanded view of this area and include the competitors' strengths and weaknesses in the assessment. It is too easy to

Opportunities versus Threats

An assessment of opportunities and threats is meaningless without an understanding of the environment in which the carrier and vendor operate.

Environment includes: demographics, politics, economics, and technology.

FIGURE 7.2 Assessment: Environment.

list a company's strengths and weaknesses and forget that a competitor's strengths and weaknesses should also be a concern. It is very useful to list strengths and weaknesses of a company and its competitors and place them side by side for review.

Strengths refer to a company's skills and available resources. Weaknesses concern a company's deficiencies. In the telecom business, weaknesses also refer to the "handcuffs" or constraints that can prohibit a carrier or vendor from performing at an optimal level. In the telecommunications world, regulation is designed to maintain a balance (sometimes one-sided) between carriers and between vendors. Regulation may differ from region to region in a single country. Legislation is also a component of the telecommunications business. Depending on your point of view legislation is either a good thing or a bad thing.

When reviewing its strengths and weaknesses, a carrier or vendor should look at specific functions or capabilities that reside within the company:

- *Marketing.* The company's level of knowledge of the customer, the market segment, and research capabilities.
- *Sales.* The company's level of knowledge of specific customers, customer segments, the ability to close the deal, to sense the needs of a customer, and to increase sales year after year.
- *Technology.* The number of patents owned by the company, the number of new patents created year after year, the technological skill levels of the company's engineers, the ability to design systems or entire networks, to operate a network, to design so that the network can easily evolve to meet customers' changing and growing needs, to repair, and to respond to emergency network operating conditions.
- *Products.* The number of products and services offered to customers. This area of concern includes the carrier's or vendor's ability to meet growth demands by the customer. Another area included is the company's ability to modify the product or service to meet the needs of the customer.

- *Materials.* Vendors must worry about the source of the materials used to build their products. Carriers today have expressed concern to vendors about the vendors' ability to purchase critical components such as computer chips in order to build their products. Any vendor should have multiple sources for specific components.
- *Finance.* The carrier's or vendor's ability to finance its growth and daily operating needs. However, there is an interesting twist to finance, because vendors will often offer their carriers financing for the equipment the carrier purchases. Financing periods of three to five years were common from the late 1990s through 2000, although long-term financing is no longer done on any large-scale. Vendors have always had to provide some financing to carriers, because carriers never pay the full balance of the contract until the equipment is installed and officially turned over for operation. Typically, vendors are paid a portion of the contract at signing, a portion at delivery, a portion at acceptance testing, and the final portion at commissioning (final acceptance).
- *Customer response.* A carrier's and vendor's customer care abilities. This area is a measure of the company's ability to communicate with the customer. Measuring whether or not a company's customer response is adequate is easy; just ask the customers what they think of the carrier's or vendor's customer response. Customers are always happy to talk about how well or badly they are being treated.
- *Management.* The strength of the management team. This includes its experience level, breadth of experience, planning skills, leadership style, and ability to manage a multitude of systems and processes.

These areas should be reviewed with an eye toward creating a list of strengths and weaknesses for a telecom company *and* its competitors. Fighting a business battle means knowing as much about the competitor as possible. Figure 7.3 is an illustration of capabilities and functions.

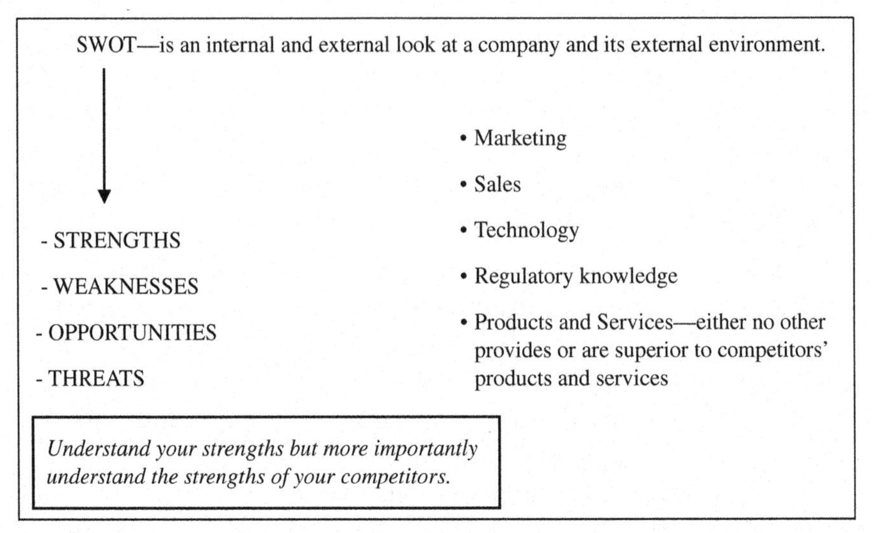

FIGURE 7.3 Telecom capabilities and functions.

OPPORTUNITIES AND THREATS. The opportunity and threat analysis refers to external factors. By now, readers will recognize that I have been describing a SWOT (Strengths, Weaknesses, Opportunities, and Threats) analysis, an internal and external look at a company and its environment.

The opportunities aspect of the strategic analysis refers to events and trends outside the company that can assist it. These beneficial external events may be new technology, a new development tool, new computing platforms, new services, a competitor's mistake, new customer needs created as a result of some new trend, or new customer purchasing trends.

Too often, an opportunity analysis fails to examine political trends. In the telecommunications business, these can mean the difference between success and failure. Opportunity analyses need to address trends in customer behavior. Customers may suddenly discover a new fad, like multicolored pagers or cellular handsets that can change the color and look of the display.

The threat aspect of the strategic analysis refers to outside events and trends that can harm the company. A threat analysis examines how changes in trends can cause potential harm. Horrible events like the September 11, 2001 World Trade

Center and Pentagon attacks will cause such environmental changes that some companies may not be able to survive. Competitors may discover a new technology that will undermine some vendor's existing product line. A carrier may be able to reduce operating costs so low that it can reduce its prices to a point no other carrier can match. A carrier price war is normal and can result in the quick demise of some carriers.

An emerging technology may cause vendors to shift future product lines to use that new technology. The technology could be a new type of computing chip that is faster than existing chips. A vendor will begin the process of installing this new chip in future shipments of the company's product, which will enable the vendor's product to stay competitive. A carrier may implement new technology at the right time in its network and as a result, acquire a large volume of new customers. All these new customers may have been another carrier's former customers. Why did customers leave? What did this emerging technology bring to customers that they liked?

A threat analysis is most effective when looked at from a scenario perspective in which the competitor can present a credible threat to the company. Scenario planning enables a telecom company to prepare for uncertainty. The telecom business is changing so rapidly that the only way to describe it in one word is "uncertainty." Scenario planning is an art with some carriers. Sometimes the only thing a carrier can do in any scenario is sit and wait for another carrier to act first. In fact, some scenarios are designed to help a carrier plan for how it will react to a competitor rather than how to take the lead. Some carriers are active followers in the business, because being the leader usually means being the first to experience all the troubles. Carriers that are successful leaders in the business are successful partially because their scenario planning is effective.

Scenario planning assumes that there is a threat and the company is in very real danger of losing. Such planning enables a company to have established plans, for a variety of business scenarios, that can be implemented quickly. A favorite term among many executives, which sums up the meaning of scenario planning, is: "What is Plan B?".

The company needs to view the marketplace from the following perspectives:

- Opportunities available to the company
- Threats posed by competitors

Often ignored but equally important is:

- The backup plan.

The backup plan is the result of scenario planning.

What is the company's escape route or alternative plan?

An Opportunity and Threat analysis has no use unless the company has a backup plan to fall on in the event the primary business battle tactics fail.

FIGURE 7.4 Opportunities and threats.

Figure 7.4 illustrates the opportunities and threats analysis.

Neither "Plan A" nor "Plan B" can be created without understanding the competitor. The competitive analysis is a component of the assessment phase and needs to examine the competitor's:

- Current market position
- Current financial health
- Operating costs, based more on educated guesswork than on factual data
- Image in the customer base
- Distribution and customer communications channels

STRATEGIC ANALYSIS

The next step in the planning process is taking the information gathered from the situation assessment and performing a strategic analysis. A stragegic analysis can be described as a discussion between managers working together and thinking through the company's problems. Strategic thinking requires people to focus on the future and the present. The issue with focusing on the present is that everyday real-world situations have an immediate impact on the company. Senior management should only be concerned with issues that affect the strategy of the company. A strategic issue or problem is a situation that affects the entire company.

Skilled senior management needs to ensure that middle management and the rest of the employee base is not distracted from important tasks. On the other hand, it also needs to be able to recognize that shifts in tasks may be necessary to accomplish the company's long-term objectives. To think through the issues surrounding any problem requires framing the issues. Before we examine how to resolve strategic issues, we need first to understand how to identify those issues. Identifying the issues or problems facing the company and determining which ones affect the direction of the company is not an easy task.

Senior managers will be asked to review the company's troubles and all of them will focus on low revenues. Each senior manager will review the situation from his/her perspective and expertise. Every senior manager will have an opinion about why the company is losing money or not making enough. Sometimes every member of the senior team will recognize the issues and develop a solution to turn the situation around. More often than not, telecom carrier and vendor executives will spend days, weeks, and sometimes months arguing over why revenues are down and what caused the company to enter the period of distress. The fact is, no single executive will want to be singled out as the primary cause or even as having a role in the company's troubles. Personal agendas within the senior team may be another factor. The first part of the strategic analysis is infighting. Reason eventually takes hold and the following process is used to identify the issues of importance. The process is comprised of a series of questions and steps in self-analysis. Essentially, a company needs to form a series of issue questions to identify the problem and eventually find the solution(s). Figure 7.5 is an illustration of the issue questions process.

Strategic Analysis leads to company management asking itself a series of questions; akin to a self-analysis.

Issue Questions—Examples:

- Are the solution(s) benefiting our customers?
- Is the outcome of the issue a symptom of the issue, the issue, or simply a remote result of the main problem?
- Did we uncover any potential obstacles?
- Did we spend enough time examining the marketplace?
- Do we understand all of the financial costs associated with our business plan?

FIGURE 7.5 Issue questions.

The senior team needs to ask itself:

- How do we know there is a problem? Did we see a reduction in orders? Did we see a drop in revenue? Did we receive complaints from customers? Are we being fined for something by the government and was it our fault? Is the outcome of the issue a symptom of the issue, the issue itself, or simply a remote result of the main problem?
- Have we found the main problem(s) or simply uncovered a series of symptoms that are related to the main problem(s)?
- Assuming we have uncovered the main problem(s), we need to test our theories. What can we do to create a test and then examine possible outcomes? The tests should be quick and easy to implement because we are attempting to identify the important vs. unimportant issues.
- Did we uncover important issues or non-issues?
- Can we implement the solution(s) at minimal cost?
- Do we need to release any employees from the payroll? If so, which ones?
- Once we implement the solution(s), will we solve our troubles for a long period or will we need to worry again next

month? (Adjusting the company's activities is a never-ending process. Competition in telecommunications is a part of the business.)

- Are the solution(s) benefitting our customers?

The worst thing that can occur is that multiple major events or issues will simultaneously affect the company. In such a situation, once the major issues are identified, management will need to classify these issues in order of importance. In telecommunications, the impact of regulatory decisions adds a level of uncertainty not found in other industries. A regulatory decision can severely affect the company, but the final decision may not be made for months or years. When the national government has made known its desire to take a certain action, the national regulatory agency issues a notice that it wishes to make a public rulemaking; before it makes any decision it would like to hear from all interested parties. By the time a national regulatory agency, like the FCC, makes its final decision, it could be years.

In a multi-issue environment, the senior management team will have created the following type of issues list:

- *Immediately impacting type.* Major issues and minor issues
- *Future/remote impacting type.* Nonissues and the potential issues

This categorization enables the management team to assign priorities to the issues in order to help resolve them. The company has a limited number of resources, which need to be managed wisely. The priority of issues is another area of contention for management team members. The experience and personal agendas of the various senior managers will generate many human sparks. When a company is in serious difficulties, the President and CEO must be present to mediate the discussion and move it forward. Such discussions work best when the senior management team actually like and respect each other. The workings of a senior team will be discussed in more detail in Chapter 8.

The simplest and most productive exercise the senior team can perform is first to identify the issues with future or remote impact. The list cannot be complete until immediate impact issues are discussed, but this is a start. The discussions to create the final list of high-impact and low-impact issues are iterative. Assuming the list is finalized and the team agrees on what the immediate and future issues are, the team must create a list of specific actions (solutions) to address the important issues first, because these important issues have an immediate impact on the company. However, this also means there must be a manageable list of issues. Figure 7.6 illustrates the types of issue priorities that need to be identified.

Creating these lists is not simple and the intensity surrounding the discussions is high. The team needs to review issues from the perspective of what is best for the company. A moderator for the meetings should be appointed. The challenge is to make sure every issue is heard and those who are involved in that issue are clearly identified. Too often, at least one senior manager will work to skirt the issue that affects him either to avoid being blamed for anything, or because of laziness. The worst thing that can happen is for a syndrome I call "you speak first, you speak loudest, or you speak the most, then you volunteer" to take hold. Sometimes issues that affect all the company's departments equally will be identified. An issue champion needs to be recognized. Since everyone is affected equally, how can this be done?

An issue champion needs to be identified.

The challenge is to make sure every issue is heard and those who are involved in that issue are clearly identified.

Examples:

- Bringing the product or service to market in a timely fashion
- Timelines
- Payroll
- Selecting Senior Team members

FIGURE 7.6 Issues priorities.

Once the issues are identified, a list of action items should be created. These action items should first be viewed from a high-level perspective, as strategic objectives. These objectives or action items are a part of the next phase of strategic planning, which can be called the decision-making phase. Creating a list of objectives is not difficult, although creating a meaningful list is. The same type of discussion to identify issues takes place when people are working to create a list of objectives.

Each objective must be set with customers in mind. Setting a financial objective without considering customers is a step closer to disaster. The customer does not care how much money you are making or that your margins are too low. The customer does not care that you need to trim costs. The customer wants quality service at a low price.

DECISION MAKING

Creating decision points means creating objectives that must be accomplished. The successful or unsuccessful completion of an objective creates a decision point for the company. I have always found that only a few key objectives need to be set and accomplished. These should have the broadest and most immediate impact on the company. When a set of objectives takes a long time to satisfy, the company runs the risk of the environment's changing and causing a new set of conditions to affect its financial outlook. Being in the middle of a self-generated change in order to meet a set of environmental changes, and then having the external environment change before you have completed your work, is a recipe for trouble.

The objectives need to be reasonable. They also need to be exciting and to articulate the vision of the company. The objectives need to energize the company's employees to want to act and accomplish something wonderful and ambitious. Employees need to feel that the decisions being made actually will move the company forward. The general employee population will look at senior management's new decisions in the light of its past ones. If the new decisions make no sense, the senior team will lose employee confidence.

Decisions need to be sound and vision oriented; their results must be measurable. Objectives need to be sound because they are articulated down the management line all the way through the general employee population. Sound objectives have understandable goals. Unsound objectives appear unachievable, and the results produce no appearance of forward movement. Do not confuse forward movement with correct movement. I have seen the wrong decisions made by a team that confused a positive result with the correct result. The senior team needs to link the results of the decisions with the next step or action in the company's life.

The worst thing a senior management team can do is generate a lot of activity that ultimately does not move the company forward in the right direction. For example, a company conducts a trial between itself and a vendor: the carrier spends a great deal of money and time on numerous telecom technology trials and does not implement any of the technology solutions. At some point, the carrier's employees start to wonder what all their hard work has meant. Poor decisions not only waste money but also damage employee morale.

SETTING OBJECTIVES. Setting objectives that lead to good decisions requires expansive thinking and flexible views. Strategic objectives cannot be set against hard and unchangeable financial measurements. Too often, senior teams believe financial goals are translatable to the employee population. The numbers mean a great deal to stockholders and even financiers, but almost nothing to employees instructed to "make it happen." Strategic objectives need to be set against the background of finance, technology, customer growth, volume sales, performance, customer care, and employee career growth.

The senior management team is working toward specific financial goals but the general employee population never has management's level of exposure. Remember, the company needs to work toward a market-driven goal. The translation of financial goals to market goals does not work for the general employee population.

A strategic objective is an objective focused on satisfying the company's overall business strategy. Strategies are broad in nature and set the direction of the company. Objectives are established to fit the strategy and not the other way around. I say this because senior managers often allow limited resources to dictate the strategy of the company; this is not a bad thing.

Two company requirements are setting achievable objectives and satisfying a vision. Even though business strategies are plans that usually point the company in the direction of a lofty goal, that goal needs to be achievable and reasonable. The goal may require $10 billion to achieve. In today's market, it is not reasonable to expect that level of financing for a startup; another way to get to the goal must be found. The strategy may require years to complete. Figure 7.7 illustrates the objective setting process.

Setting objectives against a backdrop of resources requires that the senior management team work together. This means people are negotiating and "playing nice" with each other. The first thing to do is to perform a resource analysis.

RESOURCE ANALYSIS. A resource analysis requires the company to examine how it can achieve its goals with the resources it has, or identify additional resources required to achieve those goals. Usually, the senior team will take a first pass over this matter with an eye on using existing staff and facilities. It is important for the senior team to operate within the budget, so it tries to use its current resources.

If it becomes apparent that existing employees are fully occupied and cannot split their time any more to perform

Setting objectives is an important first step in realizing the goals of the business plan.

Characteristics of Good Objectives:

- Objectives need to be realistic
- Objectives need to be focused on achieving the goals of the business plan
- Objectives need to fit the strategy
- Objectives are impacted by the resources available to the company

FIGURE 7.7 Setting objectives.

additional duties, then the next step is to determine how many more people are needed for how long. If additional people must be hired, outsourcing is an option. Outsourcing companies that provide what you need may not exist; then you need to determine if you can use existing in-house resources or hire new resources and employee training will suffice. Once the people issue is resolved, facilities and supplies can be considered.

If more space for employees is required, determine how much it will cost to obtain. If the board of directors has refused to provide the additional money to lease the additional space, then look at outsourcing. If the outsourcing option was rejected, go back to outsourcing and work even harder to find some company to do the work. If you cannot find one, go back to the board and demand the money. Figure 7.8 illustrates the resource analysis process.

If the money cannot be found, the only thing to do is modify the strategy. This can be achieved without modifying goals. The senior team needs to go through this exercise to ensure that it has thought through all the alternative paths.

What we find is that we are negotiating for resources and ultimately negotiating the strategy of the company; all of which must be must be conducted without changing the company's business plan.

IMPLEMENTATION—THE FINAL PHASE

Implementation is where "the rubber meets the road." The senior management team needs to:

- A resource analysis requires the company to examine how it can achieve its goals with the resources it has or identify additional resources needed to achieve the goals.
- Does the carrier or vendor need to outsource key functions?
- The company needs to determine what it can achieve on its own.
- The company needs to build a business with limited resources—therefore it must decide what is a priority and what is not.

FIGURE 7.8 Resource analysis.

- Create internal programs to initiate tasks
- Assign responsibilities
- Establish budgets
- Establish timelines
- Establish goals for every department

The final product of the implementation plan is not the end of the journey but the start of another phase in the life of the company. Setting the plan in motion is relatively easy, in comparison to measuring the plan's results. To some it may appear a cycle of activity. It is true that the planning process looks like a cycle, with a variety of feedback loops and checks and balances, but bear in mind that the market is always changing. The planning process is more like a never-ending journey, with lookout points along the way to check progress. Figure 7.9 is an illustration of the total strategic planning process.

IMPLEMENTATION—THE CUSTOMER VIEW

Plans are all well and good, but if the customer does not buy more or pay more for what is offered, then those plans are a failure. A carrier or vendor cannot wait for the sales results at the end of the quarter to determine if the new strategy is working.

Strategy is about asking questions and making decisions based on the information available.

The company needs to know the following in order to create a good strategy:

- Market segment and size
- Value
- Distribution
- Scope and scale

FIGURE 7.9 Strategic planning process.

If a carrier waited to take action until its (disappointing) financial results were publicly reported, it would be too late to make any course corrections to avoid another disastrous quarter.

When a carrier or vendor reports bad news for a particular quarter, it usually has figured out a full quarter before the report is made public that the news is bad. The three months prior to the report are spent trying figure out why things have gone wrong and to develop plans to emerge healthier for the next quarter. What must be done is establish a set of procedures or tools to monitor progress. The following activities need to be undertaken by the company after finalizing the strategic plan:

- Create a customer feedback loop
- Create methods to monitor customer perception
- Create methods to measure success from a customer perspective

This is far easier said than done. Measuring the pulse of the customer base is a major challenge for any carrier or vendor. It is much easier for a vendor than a carrier, because the carrier will work with a vendor it likes. The carrier will guide the vendor through its interactions.

In order to measure all vendors according to a set of standardized criteria, a carrier will articulate a process in which vendors present data to the carrier, conduct trials, and even negotiate. Carriers will also do this in order to ensure that none of the vendors pull the wool over their eyes with lots of flash. Vendors have been known to try "bait and switch" games in order to keep carriers focused on those things the vendors feel are the strong points of the product.

Carriers take control of the process and dictate how to speak with them, whom to speak to, and when to talk. By running a structured discussion process with their vendors, carriers ensure that high-pressure tactics are not used against them. Vendors will complain, but this approach does add structure to discussions that can degenerate into high-pressure sales tactics. Figure 7.10 is a depiction of how carriers can manage product discussions with vendors.

- The carrier strives to maintain control of the vendor discussion.
- The vendor strives to keep control of the discussion out of the hands of the carrier.
- The carrier needs to remember that it is always in control of the meeting because the vendor wants the carrier's money.
- The carrier has all of the leverage.
- The carrier needs to work with at least two vendors for every major system or effort.
- Competition among vendors is an excellent way of keeping any single vendor from controlling the carrier's decisions.

FIGURE 7.10 Carrier management of vendor product discussions.

The challenge faced by the carriers is with their end-users. Carriers get very few chances to make mistakes with users. Once a carrier commits an error causing the loss of a customer, the customer rarely comes back. The only time a wireless user returns to purchase service from a carrier he left is when the new carrier's service is so bad and the price is so high that the user has no choice but to return.

However, no carrier wants to treat a customer badly. First, this is wrong; it also costs a lot of money to get that customer back. Carriers employ a variety of methods to monitor customer views. Typically, they hire a third-party market research firm to conduct customer focus group meetings. Mail surveys are not very effective today, because users get bombarded with so much sales literature that the last thing a carrier wants is to spend money creating and mailing a survey that the majority of users will not answer. Besides the money spent on the survey, the carrier will just annoy the customer with the additional paper. Figure 7.11 is an illustration of how the carrier obtains input from the user.

In the past, carriers have done everything from mailing surveys stuffed with $1 bills to offering limited discounted service if users not only answer the survey but also introduce the carriers to new customers. The easiest and lowest-cost effort is the paid focus-group study. Carriers hire a firm with extensive experience in working with carriers and its own mailing list of focus-group participants. These firms have experience in structuring

- Focus groups
- Mail surveys
- Email surveys
- Marketing studies
- Telephone surveys
- Customer complaints

FIGURE 7.11 User input to carrier.

questions and analyzing the responses. The cost to the carrier is minimal in terms of money spent and mental anguish endured administering these surveys. Typically, a market research firm will give participants $50 to participate. This may sound like a great deal of money, but it is not, when compared to mailing 20 million surveys and analyzing the responses. By comparison, focus groups will survey a few thousand people. The strategy used by the carrier or vendor needs to include some kind of real-time feedback loop.

IMPLEMENTATION—THE COMPETITIVE VIEW

The development of any strategy includes data about the competitor. In the case of telecom carriers, the strategy must include a process to monitor the activities of the competitor. As I have noted, strategy setting and implementation is a dynamic process requiring companies to adjust as appropriate.

Part of the implementation process is observing how a competitor is reacting to your actions. The competitor will not know the details of your strategy, but will do everything humanly possible to figure out that strategy. Your competitor may or may not react to your market activities. You need to be careful about reacting too soon or too late to your competitor's actions, because it is possible your competitor has guessed wrong. What is important is to avoid telegraphing your strategy. Stop, think, and then act.

RESETTING THE STRATEGY

Resetting the strategy is more than adjusting the company's strategy; it is reinventing the company. My view of reinventing the company is one in which the company sets itself a new direction. The carrier is still selling telecom services, and the vendor is still selling telecom hardware and software, but now the carrier has decided to pursue a new type of telecom service, and the vendor is selling a new type of telecom technology. Corporate reinvention does not mean the company has left the business it has always served; it just means the company is serving the business differently. Consider company reinvention as the process of extending the company's products, services, and capabilities into new markets. If a carrier reinvented itself to the point that it was no longer providing telecommunications, I would call that a brand-new company, not a reinvented one. Figure 7.12 is a depiction of the reinvented company.

Why reset the strategy? The telecommunications business is so dynamic that reinvention is a necessity and the norm. The ILECs are reinventing themselves: ILECs were once considered voice telephone companies. Today they want to be Internet service providers and to provide broadband services. The ILECs are no longer plain old telephone companies—they are now telecom companies.

How does a management team know when to reset the company's strategy? The answer is when senior management

Your competitor is not sitting there happy about being number 2 because they want to be number 1 in the marketplace.

- Resetting the strategy is more than just adjusting a company's strategy: it is reinventing the company.
- Consider company reinvention the process of extending the company's products, services, and capabilities into new markets.
- Management reinvents the company in order to position the company for future challenges.

FIGURE 7.12 Telecom company reinvention.

has no choice other than shutting down. Good senior managers have one trait in common: paranoia. In this case, paranoia is not a bad thing.

Paranoia about your competitor keeps you nimble and aware, but you cannot let paranoia overwhelm you. The flip side of paranoia is complacency. A senior manager who is happy and complacent about the company's place in the market is a liability. The competitor is not happy about being Number Two, because it wants to be first in the marketplace. You happen to be in the top spot in the marketplace with the largest revenues and highest customer growth. The first thing taught in a sales department is to never, ever to get comfortable with your accomplishments. People worry about the future and that is good, but it is not the view ahead that is of concern—it is the view behind. Always be worried about your competitor's catching up (Figure 7.13).

One of the best ways to stay ahead of a competitor is to make sure that the competitor never catches up. This means monitoring not only your success, but also your competitor's success. Some carriers take a completely different view and are active followers. These carriers show up in the marketplace second, and eventually win the market by letting the leader make the marketplace mistakes first. This is a legitimate approach but means that the active follower must have resources to maintain the second-place position while waiting to take over the market.

Some telecom managers prefer to wait and watch others. The "wait and see" approach is a good tactic to use on an occasional basis. The problem with using it all the time is that you run the risk of being overwhelmed by the leader and being a very distant

One of the best ways of staying ahead of the competitor is making sure the competitor never catches up.

- Resetting corporate strategy is a powerful way of staying ahead of the competition.
- Resetting strategy is way of energizing a company to continually manage the changing environment.

FIGURE 7.13 Why reset the strategy?—competition.

second-place player. Bear in mind, U.S. telecom carriers need to encourage competition, as the Telecommunications Act of 1996 dictates. However, the leaders need not have competitors nipping at their heels.

Some managers wait until disaster is knocking at their doors before they take any kind of strategic or tactical actions. Crisis management skills are a good thing to have, but running a company using crisis management techniques is a recipe for a disaster. Some senior managers find it an adrenaline rush to handle crises, but to manage a company in this mode drains resources.

In the end, it is best to look continually in your rearview mirror. Ultimately this means that you are constantly monitoring the environment and revisiting your strategy-setting process. As illustrated in Figure 7.14, the telecom industry can be described as a dynamic business driven by technology, regulation, competitive paranoia, knowledge about the customer, crisis, and price.

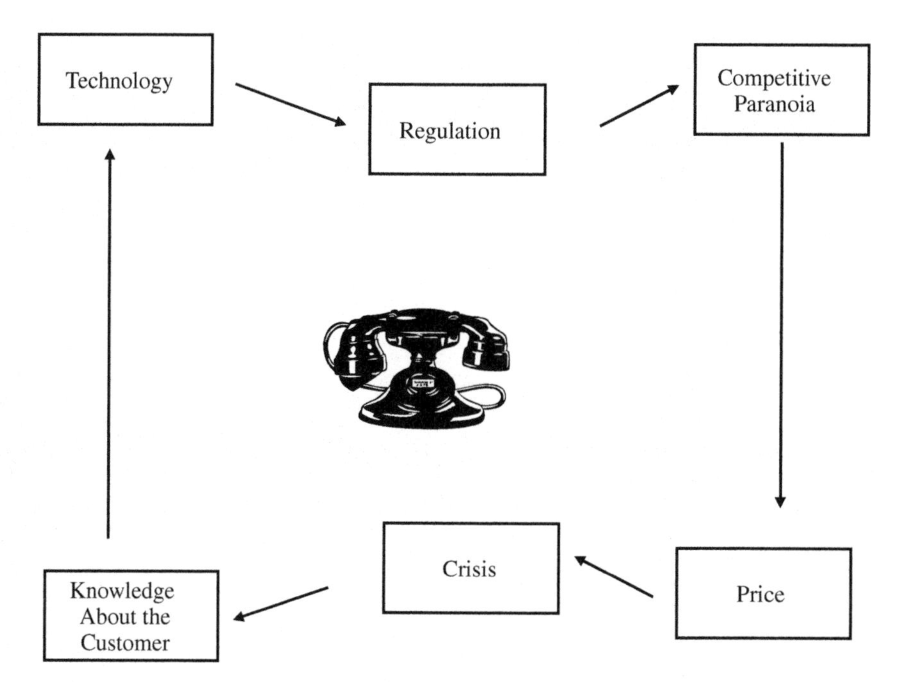

FIGURE 7.14 The drivers of the telecommunications industry.

SUMMARY

Creating a market-driven strategy involves more than just talking to and understanding customers. It requires a telecom company that can manage both a variety of internal skill sets and the impact of external forces.

We have discussed a framework in which strategies are created. We have described the factors involved in creating the strategies. Telecom carriers and vendors work diligently to craft strategies that meet marketplace needs and can react to marketplace changes. However, a process is not the only thing a telecom company needs to be successful; it also needs a management team. Processes for decision making are important, but people must make and execute the decisions.

The next chapter will examine the creation of a successful management team and the building of a market-driven company.

CHAPTER EIGHT

BUILDING THE MANAGEMENT TEAM AND BUILDING A MARKET-DRIVEN COMPANY

One thing necessary in building a company is creating a management team to lead the company. We have already discussed organizational structures and types of positions. However, we have not yet focused on the traits of a good, solid management team. A team of incredibly high performers with years of experience is still no guarantee of company success. Telecom company failures are in part due to poor execution. Business plans have been executed poorly by good managers in part because these managers were not meshing well with the other good managers. A coach once said to me that there is no "I in the word t-e-a-m." I am sure you have heard this saying many times before but it is true.

BUILDING THE MANAGEMENT TEAM

A brilliant team of managers working at cross purposes will drain company resources, spread confusion among employees and customers, ultimately demoralize the employees, and wreck the company. The team is critical. A team approach creates an environment in which the resulting brainpower is greater than the total of the individual employees' talents. The company needs to tap into the creative energies of its employees, and the first step toward this goal is to create a senior management team that works well together. Figure 8.1 depicts the concept.

A telecom carrier is only successful when all aspects of providing service to a customer are fulfilled. This means that coordination and teaming are necessary. Building a great management team starts with the President and Chief Executive Officer (CEO), who establishes a team that is a reflection of his or her personality, style, and management philosophy. A CEO who treats the staff and employee population as cattle ends up with minimal performers. Many CEOs have the good fortune to lead companies with good business plans

PRESIDENT + CEO

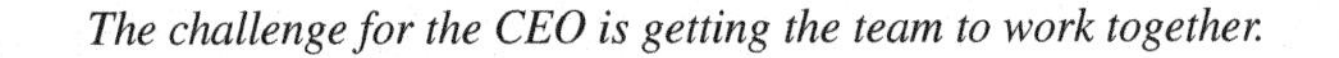

If the CEO does not have his or her team operating with these three factors in force, than the company will have severe management issues.

HONESTY

RESPECT

OPENNESS

FIGURE 8.1 Teaming.

and much financial support. These companies attract great talent. The challenge for the CEO is getting the team to work together. Every CEO has his or her own style of managing and communicating.

The only comment I make about style is that it is vital to respect everyone in the company. You cannot be a CEO of an empty company. Some CEOs and senior managers are good at acting their parts and communicating what is required of them. It would be decent for the senior management to express respect sincerely, but if employees can live with the acting, that will do.

If the CEO is dealing with experienced managers, there should be no need to communicate the importance of working together. However, when personalities and egos get in the way, which often happens, the CEO needs to communicate cooperation among his or her direct reports. The CEO should work to establish the guideposts of respect, honesty, and openness among the senior management.

Respect for each other among senior managers is one important factor. Respect of senior managers for the general employee population is another factor. Respect is a necessity. Imagine working with someone who thinks you are an idiot, too stupid to work your way out of the corner of a room. The person you are working with may say all the right things, but his or her actions say something else. Actions speak louder than words. Memories are selective, but actions leave their mark. Senior teams have disintegrated because of a lack of respect. Employees will recognize when a senior manager thinks they are fools. Employees who do not like their bosses do just enough of their jobs to survive.

Honesty among the senior team has been found lacking in some companies. Honesty is important. Each member of the senior management team must express themselves without fear of retribution. There is no call for lack of professionalism or rudeness. However, if senior management cannot speak about issues (with one another) without being ridiculed or treated as enemies, then the CEO will never be able to harness the positive energy and ideas of the senior team. If one vice president needs to worry about being stabbed in the back over some issue by another vice president, what kind of effective management team can a company have? The answer is none at all.

Openness means being open to new ideas and the ideas of others. The team needs to discuss issues openly without fear and reprisal.

The factors I have noted will only have a positive effect on the way the company is managed if *all* of the factors are operating. The team must respect one another and others, and be honest and open with each other. If the CEO does not have his or her team operating with these three factors in force, then the company will have severe management difficulties. This point is illustrated in Figure 8.2.

Without respect, honesty, and openness, there is no way for the CEO to establish a mutually reinforcing environment for successful management. The key to creating a successful team is making sure the team intentionally follows these rules. The challenge is one cannot make the team do something it will not support. Therefore, the CEO must hire senior managers who believe in him or her and have the same basic philosophy of life and management. If the senior team is not closely aligned with the CEO in style and thinking, the CEO will not be able to run the company. The CEO needs to hire people he or she can trust. Trust is earned; it cannot be bought or paid for.

If the CEO does not personally know the person they are hiring, than that CEO needs to get to know that person quickly. The safest course for the CEO and the company is to hire at

The CEO needs to hire people he or she can trust.

- Trust is earned.
- Trust cannot be bought or paid for.

- Senior management is critical to the overall health of the company.
- The senior management team needs to work with each other—respect, honesty, and openness.

FIGURE 8.2 Respect, honesty, and openness.

least one loyal senior manager who has worked with the CEO before.

ROLES AND RESPONSIBILITIES

Senior management teams are comprised of individuals with a variety of skill sets. In a carrier, the management team members have a strong understanding of each other's responsibilities. This also includes respecting each other's turf.

In a well-run senior staff meeting there will be an exchange of ideas. However, only one senior executive will be held accountable for implementing the final decision. At the end of the meeting, every other senior executive must back off and let the responsible vice president do his or her job. There must be recognition of each other's experience and knowledge; however, there must also be recognition of roles and responsibilities.

BUILDING A MARKET-DRIVEN COMPANY

How do you know your company is market driven? The answer is when your customers say they are happy and are paying you money. A market-driven company has a number of embedded customer-oriented characteristics, which take on the nature of a philosophy. The employees breathe this philosophy as if it were air.

The employees view this philosophy central to their being; core to everything that makes them who they are. It is what drives them to succeed for the company. Figure 8.3 illustrates the philosophy behind a market-driven company. This philosophy is:

- Bringing value to the customer
- Making the customer the first consideration
- Acknowledging that the customer is always right
- Doing everything humanly possible to bring value to customers

- The customer comes first.
- The customer is always right
- Meeting the customer's needs in a timely fashion
- The customer has the money
- The carrier and vendor need the money
- The customer will only pay us if they are happy and their needs are met

The customer is the focus.

FIGURE 8.3 The philosophy of a market-driven carrier/vendor.

The mission of the carrier is to bring happiness and value to its customers. Every employee must remember the following:

- The customer is my friend
- The customer is more important than I am
- The customer has the money
- We need the money
- The customer will only pay us if he or she is happy
- Dissatisfied customers means no money and I will be out of a job

Does this sound silly? Think again. A staff should be thinking this or they will make a mistake like forgetting the customer is king or queen. Many vice presidents of sales and CEOs have encountered situations where, for just one moment, their staffs forgot the customer came first and blew a big deal.

CHARACTERISTICS OF A MARKET-DRIVEN COMPANY

As silly as the above list may have sounded, its points are true: they describe a series of characteristics found in any market-driven company. The characteristics include:

- A common set of beliefs and values
- Understanding of the customers' needs, behavior, and attitudes
- Commitment to the customer
- Encouragement of market-driven behavior in the company
- Aligning the company's resources to meet the needs of the marketplace

Figure 8.4 is a depiction of the characteristics of a market-driven carrier or vendor.

COMMITMENT TO BELIEFS AND VALUES

Market-driven companies must be completely focused on the customer. They need to worry about meeting sales projections, but the best way to ensure the company is meeting its "numbers" is by focusing on the customer. A single theme must run in the minds of every carrier and vendor employee—the customer comes first.

This means:

- High quality of service or products for the customer
- Concern that the customer's needs have been met
- Observing competitors—look for new opportunities and copy a good idea when you see one
- If an idea is bad, a service has worn out its usefulness, or a product is no longer of value, do not hesitate to get rid of it.

- Common set of beliefs and values
- Understanding of the customer's needs, behaviors, and attitudes
- Commitment to the customer
- Encourage market driven behavior in the company
- Align the company's resources to meet the needs of the marketplace

FIGURE 8.4 Characteristics of a market-driven carrier and vendor.

A market-driven company is prepared to shut down a product or service line if necessary.

Market-driven companies are not afraid to change; they are only concerned about keeping a customer's business.

CUSTOMER NEEDS, BEHAVIOR, AND ATTITUDES

Market-driven companies are constantly listening to customers. Market driven companies track purchasing patterns. Market driven companies constantly survey their customers. Market-driven companies constantly collect data about their current and potential customers. They analyze the data to mine valuable information about the marketplace.

Needs, behavior, and attitudes are all factors that affect the way a customer buys telecommunications services and products. All these factors are subjective in nature. There is no right or wrong in regard to customers' needs, behavior, and attitudes. A customer's telecommunications needs can be categorized as follows:

- Personal
- Business
- Emergency
- Entertainment
- Educational
- Travel

A customer's behavior and attitudes can be influenced by the following factors:

- Professional situations
- Home situations
- Weather
- Politics

- Personal cash liquidity

A market-driven company needs to sense the state of the marketplace in order to serve that marketplace.

COMMITMENT TO THE CUSTOMER

This should be an obvious necessity. Commitment to the customer means a deep-rooted belief within the telecom business that everything must be done to serve the customer. If the carrier has priced a service too high, then it needs to reprice that service.

Of course, if the customer is willing to pay a high price for the service, then the company should leave well enough alone. This may sound odd in light of keeping the customer happy, but it is not. I have stated that the carrier or vendor needs to keep the customer happy; but it cannot be foolish and lose money. Commitment to the customer is a fine line along which carriers and vendors constantly balance themselves. Carriers and vendors need to be committed to customers, but generating a net profit is still necessary.

However, fulfilling a company's commitment to customers may mean spending a great deal of money doing market research or operating a customer care center. Commitment translates into dollars spent on understanding customers' needs, behavior, and attitudes. Travel dollars are another way of showing commitment to the marketplace and customers. Sales and marketing people will often need to travel to meet customers. Times are bad for a telecom vendor that reduces its travel and entertainment budget, or for a telecom carrier when it reduces its advertising and market research budgets.

How committed to the customer is your carrier or vendor? What hoops will the carrier or vendor jump through for your business? How much money is the carrier or vendor willing to spend in order to keep the business and make more money?

ENCOURAGING AND MAINTAINING MARKET-DRIVEN BEHAVIOR

Words and actions are important to a company. Carriers and vendors need to encourage market-driven behavior and maintain the energy and excitement surrounding such behavior. Market-driven behavior needs to be rewarded so that it is encouraged to grow and flourish.

As I noted, words and actions galvanize a company. The telecom service provisioning business is filled with professionals passionate about serving customers. Even though the passion comes naturally for most telecom professionals, these people need to be rewarded for their efforts. Money is usually the way a company rewards positive behavior. Sometimes awards like certificates or plaques are used to acknowledge people's efforts. Acknowledgments should be publicly made. Public recognition of jobs well done goes a long way toward building loyalty and excitement among employees. Figure 8.5 is an illustration of points.

ALIGNING COMPANY RESOURCES

Aligning corporate resources to ensure customers' needs are being met is a costly and intensive way to demonstrate commitment; however, it has been done, as we have noted throughout the book. Carriers are organized to optimize their resources and meet customer needs. When the marketplace changes suf-

- Often the carrier and vendor will take steps to organize its activities and its structure to ensure that the customers are being served properly.
- Cooperation means a happy user.

Public recognition of jobs well done go a long way towards building loyalty and excitement among employees.

Money is usually the way a company rewards positive behavior.

FIGURE 8.5 Encouraging and awarding market-driven behavior.

ficiently, carriers and vendors often take steps to organize their activities and structures to ensure that the customers are being served properly.

Market-driven companies run their internal organizations so that they all focus their creative energies, activities, and decisions around opportunities and customer issues. Total reorganizations are not always necessary. Sometimes a readjustment is all that is needed.

As I noted in Figure 8.6, what is important is that the company's resources are focused on the marketplace. Telecom carriers and vendors have a difficult time keeping abreast of marketplace changes. Since technology plays such an important role in telecommunications, a heretofore unseen technology and resulting product could cause unanticipated and massive changes in the marketplace. The Internet, personal data assistant (PDA), wireless Internet, and fiber optics caused such a stir in the marketplace that it took many carriers years to understand how they could play a role in occupying the business space created by these products. Some carriers even closed their doors.

Realigning a carrier's resources to meet new customer needs is a major task. Therefore, it is important to have an organizational structure in place that lends itself to flexibility. The operating company model known as *centralized control* of *decentralized operations* is the best model to establish in an

Focusing and aligning resources is achieved by:

- A goal that is clearly understood amongst employees.
- Company cooperation across all departments is necessary.
- Internal project management.
- A senior management team that manages and leads well.

Why Focus Resources?

- The alternative is a company that lacks focus and cannot meet customers' needs properly.

FIGURE 8.6 Focusing company resources.

environment of change. The Sloan model is a balance of the best of both worlds. Large companies today are global, with multiple operations and many thousands of employees.

SUMMARY

The different company types in the telecom space include:

- Makers of hardware
- Makers of software
- Providers of service to the hardware and software companies
- Providers of service (carriers/service providers) to the user
- Providers of service to other providers of service

Telecom equipment is complex. The corollary is that the business of creating the product is just as complex. Service provisioning is as complex a business as one can be in, especially since the service is so technology dependent. Technology is constantly changing and evolving. Change is a component of the telecom business.

Managing change is an art that cannot be properly taught in a school. It is possible to learn the mechanisms and steps of managing change, but executing change management is best learned in the real world. To run a carrier or vendor well requires a good, solidly dependable senior management team.

A winning management team needs to listen to employees and to customers. It must embrace the attributes of respect, honesty, and openness if it expects to function properly.

As Figure 8.7 illustrates, the telecom industry is going through a period of change, not all of it pleasant. Uncertain economic times have destroyed some fine carriers and vendors. However, one thing we can be certain of is that people will always need to communicate and will need telecom services ranging from plain old voice to Internet service.

As I have noted, strategy means asking questions and making decisions based on available information. Carriers and ven-

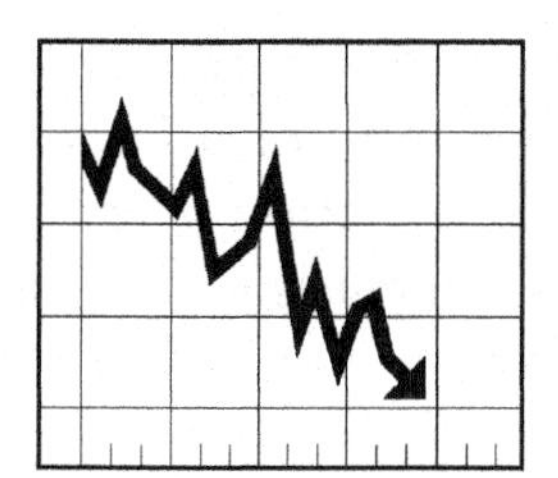

FIGURE 8.7 Telecommunications and change.

dors need to develop a variety of scenarios in which a series of "what if" questions are asked and a series of possible answers are created. The "what if" questions I mentioned serve as the basis for the activity known as scenario planning.

Strategic planning is a dynamic process. There is a goal, but the plan to get to that goal may need to change as time and circumstances change. Creating a market-driven strategy requires a telecom company to manage both a variety of internal skill sets and the impact of external forces. At the end of the day, we need a good management team and a market focus.

The year 2001 and 2002 can be characterized as a period of recession and stabelization. The telecommunications industry has been rocked by enormous change, all of which as been painful for those in the industry. The telecommunications industry is effectively right-setting itself. Too many telecom (including Internet) startups have failed. Unfortunately, a combination of poor execution and poor business plans fed the market frenzy to make quick money.

The industry is at a critical juncture of change. Broadband services are at the brink of becoming a commercial success. However, success of broadband wil be the result of many new consumer applications hitting the marketplace. Opportunity in telecommunications has always been about the succesful commercialization of many different services. The Internet is key to the succcess of the emerging broadband market.

APPENDIX A

ACRONYMS, DEFINITIONS, AND TERMINOLOGY

Access link (A-link) Used for Switch-STP signaling connections and STP-HLR.

Access Network Portion of a public switched network that connects access nodes to individual subscribers.

Access Service Area A geographic area established for the provision and administration of communications service. An access service area encompasses one or more exchanges, where an exchange is a unit of the communications network consisting of the distribution facilities within the area served by one or more End Offices, together with the associated facilities used in furnishing communications service within the area.

Access Tandem (AT) **1.** A switching system that concentrates and distributes traffic for interLATA traffic originating or terminating within a LATA (local access and transport area). An access tandem can also provide equal access for nonconforming end offices. **2.** An EC switching system that provides a traffic concentration and distribution function for interexchange traffic originating or terminating within an access service area.

Address Signals These convey call destination information or the digits dialed by the calling party. There are several types of address signaling: Dial Pulse, DTMF (Dual-Tone Multi-Frequency) and MF (Multi-Frequency).

Advanced Intelligent Network (AIN) A telecommunications network capable of providing advanced services that can be used on wireless and wired networks through the use of centralized databases that can provide call processing and routing.

Alerting Signals These alert the user of an incoming call. In other words, this is the ringing the user hears; also known as power ringing in the wireline world.

Amateur satellite service Satellite service for radio amateur use in carrying out technical investigation and learning about intercommunications.

American National Standards Institute (ANSI) A nonprofit organization that coordinates voluntary standards activities in the United States. The institute represents the United States in two major telecommunication organizations: the International Standards Organization (ISO) and the International Electrotechnical Commission (IEC).

American Standard Code for Information Interchange (ASCII) A widely accepted standard for data communications that uses a 7-bit digital character code to represent text and numeric characters. When companies use ASCII as a standard, they are able to transfer text messages between computers and display devices regardless of the device manufacturer.

Analog Signal A signal that is modified in a constant fashion, such as voice or data.

Answer Supervision/Answer Signal The signaling state of a circuit that indicates idle or busy, on-hook or off-hook status. The signaling state may be indicated in various ways, depending on the signaling system.

Area Code A three-digit number that identifies the home area of a telephone. In North America, the NPA (Numbering Plan Area) represents the Area Code.

Asymmetric Digital Subscriber Line (ADSL) Modems attached to twisted pair copper wiring that transmit from 1.5 Mbps to 9 Mbps downstream (to the subscriber) and from 16 Kbps to 800 Kbps upstream, depending on line distance.

Attenuation The decrease in power of a signal, energy, light, or radio signal. This decrease can occur either as a fractional value or completely.

Authentication A process during which information is exchanged between a communications device (typically a mobile phone) and a communications network that allows the carrier or network operator to confirm the true identity of the unit. This inhibits fraudulent use of the mobile unit. The process of user identity confirmation. Identity confirmation can involve checking handset or terminal device identity by interpreting "secret" keys or data messages. If the data keys or data messages have been altered or do not show specific format, the call will not be completed.

Automatic Number Identification (ANI) The number providing the billing number of the line or trunk that originated a call. It is a charge number used to support exchange access and billing. This number may or may not be identical to the Calling Party Number.

Backbone A common distribution channel that carries analog or digital telecommunications signals for many users. Also, the central distribution cable from an interface.

Bandwidth The width of a radio channel (in Hertz) that can be modulated to transfer information.

Basic Rate Interface (BRI) In ISDN, the network interface that provides 144 kb/s information transfer as defined in ANSI Standard T1.607.

Bearer In the communications industry, a transmission channel used to carry data. In ISDN, there are 64 kbps bearer channels used.

Billing System A system that records the occurrence of a call or some event, the identity of the originating party, the identity of the destination party, and the time length of the call. This system must also process the data for the rendering of a bill to the subscriber.

Bit The smallest part of a digital signal, typically called a data bit. A bit typically can assume two levels; either a zero (0) or a one (1).

Bit error rate (BER) A measurement used to determine the quality of a digital transmission channel. BER measures the ratio of bits received in error compared to the total number of bits transmitted.

Bridge A data communications device that connects two or more networks and transmits information between the networks.

Broadband Typically refers to voice, data, and/or video communications at rates greater than wideband communications rates (1.544 Mbit/s).

Broadband Integrated Services Digital Network (B-ISDN) A digital network with ATM switching operating at data rates in excess of 1.544 or 2.048 Mbps. ATM enables transport and switching of voice, data, image, and video over the same infrastructure.

Broadcast messaging Messaging between users of telecommunications networks. Typically, broadcast messaging is between one user and many users.

Broadcasting satellite service Allows sound and visuals to be received by individuals or communications via satellite.

Broadcasting Service Service in which the transmissions are intended for direct reception by some consumer group, generally the public at large.

Bursty data Data rates that fluctuate widely with no predictable pattern.

Busy Hour A time-consistent hour in a specific measurement period when the total load offered to a group of trunks, a network of trunks, or a switching system is greater than at any other time-consistent hour during the same measurement period.

Cable Television (CATV) *See Community Access Television.*

Call Delivery (CD) A call routing process that permits a subscriber to receive calls to his directory number while roaming.

Call Detail Recording (CDR) Telecommunications system's ability to collect and record detailed information on all outgoing and incoming calls.

Call Forwarding-Busy (CFB) A call routing service that permits a called subscriber to have the system send incoming calls addressed to the his directory number to another directory number (forward-to number) or a designated voice mail box, when the subscriber is engaged in a call or service.

Call Forwarding-Default (CFD) A call routing service that permits a called subscriber to send incoming calls addressed to his directory number, designated voice mail box, or another directory number (forward-to number) when the subscriber is engaged in a call, does not respond to paging, does not answer the call within a specified period after being alerted, or is otherwise inaccessible (including no paging response, the subscriber's location is not known, the subscriber is reported as inactive).

Call Forwarding-No Answer (CFNA) A call routing service that permits a called subscriber to have the system send incoming calls addressed to his directory number to another directory number (forward-to number), or to his subscriber's designated voice mail box, when the subscriber fails to answer, or is otherwise inaccessible (including no paging response, the subscriber's location is not known, the subscriber is reported as inactive, Call Delivery is not active for a roaming subscriber, Do Not Disturb is active, etc.).

Call Forwarding-Unconditional (CFU) A call routing service that permits a called subscriber to send incoming calls addressed to his directory number to another directory number (forward-to number) or to his designated voice mail box. If this feature is active, calls are forwarded regardless of the condition of the termination.

Call processing Steps that occur during the duration of a call. These steps are typically associated with the routing and control of the call.

Call Progress Signals Voice band tones and announcements used to inform a calling end-user or operator of the progress or disposition of a call. Examples include busy tone, voice announcement, special information tones, and audible ringing. The tones or announcements can be used after a number has been dialed and until the called telephone is answered or the attempt is abandoned.

Call Routing In circuit switching, the process of determining the path of a call from point of origination to point of destination.

Call Transfer (CT) A call-handling feature that transfers a call from one station or extension. Call transfer service enables a subscriber to transfer an in-progress established call to a third party. The call to be transferred may be an incoming or outgoing call.

Call Waiting (CW) A telephone call-processing feature that notifies a telephone user that another incoming call is waiting to be answered. This is typically provided by a brief tone not heard by the other callers. Some advanced telephones (such as GSM mobile telephones) are capable of displaying the phone number of the waiting call. CW is a call routing service that provides notification to a controlling subscriber of an incoming call while his call is in the two-way state. Subsequently, the controlling subscriber can either answer or ignore the incoming call. If the controlling subscriber answers the second call, he may alternate between the two calls.

Calling Name Identification Presentation (CNaIP) A call processing service that provides the name identification of the calling party to the called subscriber.

Calling Name Identification Restriction (CNaIR) A call processing service that restricts the presentation of the calling party's name to the called subscriber.

Calling Number Identification (CNI) A feature that allows a telephone customer to view the telephone number of the person who is calling. A related service, Calling Number Identification Restriction (CNIR), allows the caller to inhibit the display of his telephone number when placing a call.

Calling Number Identification Presentation (CNIP) A call routing service that provides the number identification of the calling party to the called subscriber. One or two numbers may be presented to identify the calling party.

Calling Number Identification Restriction (CNIR) A call processing service that restricts presentation of that subscriber's Calling Number Identification (CNI) to the called party.

Calling Party Number (CPN) In the Integrated Services Digital Network (ISDN) Q.931 and Signaling System 7 protocols, an information element that identifies the number of an originating party. A set of digits and related indicators that provide numbering information related to the calling party.

CCS (Centum [Hundred] Call Seconds) A measurement of telephone usage traffic used to express the average number of calls in progress or the average number of devices used.

Cell Site A transmitter-receiver tower, operated by a wireless carrier (typically cellular or PCS), through which radio links are established between a wireless system and mobile and portable units.

Central Office A term used to describe local switches used by the local telephone companies. The term is also synonymous with end office.

Centrex A service for business customers that shifts the functions usually associated with a private branch exchange (PBX) on a customer's premises to a central-office switching system.

Channel Reuse The practice of independently using radio channels that have the same radio frequency to cover different coverage areas.

Coaxial Cable A transmission line consisting of an inner conductor surrounded first by an insulating material and then by an outer conductor, either solid or braided. The mechanical dimensions of the cable determine its characteristic impedance.

Common Channel Signaling (CCS) Also known as *out-of-band* signaling; describes a scheme in which the content of the call is separated from the information used to set up the call (signaling information).

Community Access Television (CATV) Also known as Cable TV. A popular form of sending entertainment programming into households via a physical transmission system (e.g, coaxial cable, DS1, etc.). Years ago, CATV enabled homes in remote areas to receive limited television programming when wireless transmission signals could not reach such areas.

Competitive Access Provider (CAP) An alternative local exchange carrier that competes with the existing incumbent and dominant local exchange carrier.

Competitive Local Exchange Carrier (CLEC) A description of the new carriers (as encouraged by the

Telecommunications Act of 1996) that are competing in the local loop marketplace.

Conference Calling (CC) A call routing service that provides a subscriber with the ability to have a multiconnection call, i.e., a simultaneous communication between three or more parties (conferees).

Control Signals Used for special auxiliary functions that are beyond a service provider's network. These signals communicate information that enables or disables certain types of calls. One example would be call barring.

Convergence Addresses the technical and business aspects of integration of technology and business.

Country Code A one-, two-, or three-digit number that identifies a country or numbering plan to which international calls are routed. The first digit is always a world zone number. Additional digits define a specific geographic area, usually a specific country.

Customer Care System A customer profile database system used to support customer complaints, add new subscribers, remove subscribers, and contain customer profile information.

Customer Premises Equipment (CPE) All telecommunications terminal equipment located on the customer's premises, including telephone sets, private branch exchanges (PBXs), data terminals, and customer-owned coin-operated telephones.

Data Compression A technique for encoding information so that fewer data bits of information are required to represent a given amount of data. Compression allows the transmission of more data over a given amount of time and circuit capacity. It also reduces the amount of memory required for data storage.

Data Terminal Equipment (DTE) In a data communications network, the data source, such as a computer, and the data sink, such as an optical storage device.

Data Warehouse An information management service that stores, analyzes, and processes information derived from transaction systems.

Database A collection of interrelated data stored in computer memory with a minimum of redundancy. Database information held in a computer-accessed memory usually is subdivided into pages, with each page accessible to all users unless it belongs to a closed user group.

De-multiplexing A process applied to a multiplexed signal for recovering signals combined within it and for restoring these individual signals.

Dedicated Circuits Circuits designated for a specific use or customer.

Dial-Line A two-wire, line-side connection from a LEC end office. This connection is like those used for business and residential lines. It may be used on a one-way or two-way basis. Dial-Line connections enable the MSC to access any valid telephone number.

Dialing Parity A company that is not an affiliate of a local phone company is able to provide phone services so that customers can route their calls automatically without the use of an access code.

Digital Signal A signal with a limited number of discrete states, usually two. In contrast, an analog signal varies continuously and has an infinite number of states.

Digital Signal 0 (DS0) A 64-kbps digital representation of voice.

Digital Signal 1 (DS1) Twenty-four voice channels packed into a 193-bit frame and transmitted at 1.544 Mbps. The unframed version, or payload, is 192 bits at a rate of 1.536 Mbps.

Digital Signal 2 (DS2) Four T1 frames packed into a higher-level frame transmitted at 6.312 Mbps.

Digital Signal 3 (DS3) Twenty-eight T1 frames packed into a higher-level frame transmitted at 44.736 Mbps.

Digital Subscriber Line (DSL) A two-wire, full-duplex transmission system that transports user data between a customer's premises and a digital switching system or remote terminal at 144 kbit/s.

Digital Television A device that receives broadcast digitally formatted television signals. Such devices display programming with far higher resolution than so current television sets.

Direct Distance Dialing (DDD) A telephone service that lets a user dial a long distance call without operator assistance. This is a common feature today, but years ago it was considered a special capability for telephone switches.

Direct Inward Dialing (DID) A private branch exchange (PBX) or Centrex feature that enables completion of an incoming call directly to an extension station without operator assistance. Direct inward dialing is also used in voice mail and radio paging systems.

Disconnect Signal An on-hook signal indicating the connection is being cleared. The signal responding to a disconnect signal, but applied in the direction opposite to the direction of propagation of the disconnect signal may also be considered a disconnect signal.

Distinctive Ringing A feature that enables a subscriber's multiplicity of numbers (on a single line) to be identified by different ringing patterns.

Do Not Disturb (DND) A call-processing feature that prevents a called subscriber from receiving calls. When this feature is active, no incoming calls will be offered to the subscriber. DND also blocks other alerting, such as the Call Forwarding-Unconditional abbreviated (or reminder) alerting and Message Waiting Notification alerting. DND makes the subscriber inaccessible for call delivery.

Domain Name The unique name that identifies an Internet site.

Downlink The portion of a communication link used for transmission of signals from a satellite or airborne platform to a terrestrial terminal.

Dual Tone Multi-Frequency (DTMF) A means of signaling that uses a simultaneous combination of one of a lower group of frequencies and one of a higher group of frequencies to represent each digit or character. These frequencies are tones.

E Carrier A 10-MHz wireless carrier also known as a Personal Communications Service (PCS) carrier.

E1 European basic multiplex rate that packs 30 voice channels into a 256-bit frame transmitted at 2.048 Mbps.

Earth Exploration Satellite Service Service that allows observation of the Earth for various purposes, such as for weather or geological information.

Electronic Mail (E-mail) Messages, usually text, sent from one person to another via computer.

Emergency Services Access Point An emergency services network element that is responsible for answering emergency calls.

Encryption A process of a protecting voice or data information from being obtained by unauthorized users. Encryption involves the use of a data-processing algorithm (formula program) that employs one or more secret keys used by both the sender and receiver of the information to encrypt and decrypt the information. Without the encryption algorithm and key, unauthorized listeners cannot decode the message.

End Office (EO) An EO switching system that terminates station loops and connects the loops to each other and to trunks.

Engineered Capacity The highest load level at which service objectives are met for a trunk group or a switching system.

Equal Access An exchange carrier service that gives a customer an equal choice of trunk-side access to public switched interexchange carrier telephone networks in terms of such items as dialing plan and transmission quality. Another term for this is Feature Group D.

Erlang Amount of voice connection time with reference to one hour. For example, a six-minute call is .1 Erlang.

Ethernet A transmission protocol for packet-switched local area networks (LANs). Ethernet is a registered trademark of Xerox Corporation.

Exchange The typical shorthand term used to identify a telephone switching center.

Exchange Access The offering of access to telephone exchange services or facilities for the purpose of the origination or termination of telephone toll services.

Exchange Carrier (EC) A telephone company, generally regulated by a state regulatory body, that provides local (intraLATA) telecommunications services.

Facilities The transmission parts (elements) of a service provider. Sometimes used in more general terms to describe buildings and utilities.

Facsimile Also known as *fax*. A form of telegraphy that supports the transmission of fixed images to some form of fixed media, like paper.

Fax Mail (FxM) A communications service that provides a subscriber with fax mail services, which include the ability to store faxes received and forward faxes recorded, based on a number of subscriber-set parameters.

Fiber to the Home (FTTH) A communication network in which an optical fiber runs from telephone switch to the subscriber's premises or home.

Fiber to the Neighborhood (FTTN) A communications network in which an optical fiber runs from a switch to a neighborhood of homes.

Firewall A physical and electronic method of protecting computers from outside attack. This protection involves hardware and software.

Fixed Satellite Service A radio communication service that addresses communication between earth stations at specified fixed points via one or more satellites (e.g., Intelsat).

Flat Rate Price-setting principles for a service provider who wishes to charge the same fee for calls regardless of the number of calls made or the duration of each call. Usually flat rates are single monthly or periodic charges.

Flexible Alerting (FA) A call-processing service that causes a call to a pilot directory number to branch into several legs to alert several termination addresses simultaneously. The mobile telephones in the group may be alerted using distinctive alerting. Additional calls may be delivered to the FA pilot directory number at any time. The first leg to be answered is connected to the calling party. The other call legs are abandoned.

Foreign Exchange (FX or FEX) Telecommunications trunk lines that connect directly to a foreign telephone company's switching system (exchange).

Four-Wire Circuits A physical path in which four wires to the terminal equipment are represented. This path allows for simultaneous transmission and reception. Two wires are used for transmitting in one direction and the other two wires are used for transmitting in the other direction.

Fractional T1 Fractional T1 refers to data transmission speeds between 56 Kbps and 1.544 Mbps in a single full T1.

Frame Relay An access standard defined by the standards body called ITU. Frame relay services are telecom services that employ a form of packet switching analogous to a streamlined version of X.25 networks. The data packets are in the form of "frames," which are variable in length.

Frame Switching Devices that forward information frames based on the frame's layer 2 address. Frame switching can be done in two ways: via cut-through switching or via store-and-forward information switching.

Frequency Division Multiplexing (FDM) A technique in which the available transmission bandwidth of a facility is divided by frequency into narrower bands. Each band is used for a separate voice or data transmission channel. In this manner, multiple conversations can take place over a single transmission facility.

Full Duplex Transferring of voice/data in both directions at the same time. This becomes confusing in a TDMA system because information is reconstructed to allow transfer of voice information in both directions at the same time, although actual transmission does not occur simultaneously.

Gateway A device or facility that enables information to be exchanged between two dissimilar computer systems or data networks. A gateway reformats data and protocols in such a way that the two systems or networks can communicate.

Global Title In the Signaling System 7 (SS7) protocol, an address, such as customer-dialed digits, that does not contain explicit information to enable routing in a signaling network and, therefore, requires the signaling connection control part translation function.

Global Title Translation (GTT) In SS7, a procedure that translates a global title into a known address that allows message routing in the signaling network.

GOS (Grade of Service) An estimate of customer satisfaction with a particular aspect of service. It also refers to the probability that a call will fail due to the unavailability of links or circuits.

Half Duplex The ability to transfer voice or data information in either direction between communications devices but not at the same time. The information is transmitted on one frequency and received on another.

Handoff The process of reassigning subscriber handsets to specific radio channels as the handsets move from cell site to cell site.

Hard Handoff A "break-before-make" form of call handoff between radio channels. In this scenario, the mobile handset temporarily (time measured in milliseconds) disconnects from the network as it changes channels. The radio protocols AMPS, TDMA, and GSM support only hard handoff.

High Data Rate Digital Subscribe Line (HDSL) Modems on either end of one or more twisted pair wires that deliver T1 or E1 speeds. At present T1 requires two lines and E1 requires three.

High Level Data Link Control (HDLC) A bit-oriented communications protocol in which control codes differ according to their bit positions and patterns.

High Usage Trunk Group A transmission facility used only for routing large volumes of traffic to a single point or set of points.

Hybrid Fiber Coax A system (usually CATV) where fiber is run to a distribution point close to the subscriber and then the signal is converted to run to the subscriber's premises over coaxial cable.

Inband Signaling Signaling in which the frequencies or time slots used to carry the signals are within the bandwidth of the information channel.

Incumbent Local Exchange Carrier (ILEC) A telephone service carrier that operated in the local loop market prior to the Telecommunications Act of 1996 and the divestiture of the AT&T Bell system. In other words, the local telephone company most people grew up with.

Information Service The offering of a capability for generating, acquiring, storing, transforming, processing, retrieving, utilizing, or making available information via telecommunications; it includes electronic publishing, but no use of any such capability for the management, control, or operation of a telecom system or the management of a telecom service.

Integrated Services Digital Network (ISDN) A structured all-digital telephone network system developed to replace (upgrade) existing analog telephone networks. The ISDN network supports advanced telecommunications services and defined universal standard interfaces used in wireless and wired communications systems.

Interconnection The connection of telephone equipment or communications systems to the facilities of another network. The FCC regulates interconnection of systems to the public switched telephone network.

Interexchange Carrier (IXC) A carrier company in the United States, including Puerto Rico and the Virgin Islands, that is engaged in the provision of interLATA, interstate, and/or international telecommunications over its own transmission facilities or facilities provided by other interexchange carriers.

Interexchange Trunks Transmission facilities used to support communications between switching centers (LEC switches) and the interexchange carrier (IXC).

Interlata Service Telecommunications between a point located in a local access and transport area and a point located outside such area.

International Carrier (INC) A carrier authorized to provide interexchange communications services outside World Zone 1 using the international dialing plan; however, the carrier has the option of providing service to World Zone 1 points outside the 48 contiguous states of the United States.

International Gateway Facilities (IGFs) Transmission facilities used by international gateway switches. An international gateway switch is used by the long distance (interexchange) carrier to interface with international telecommunications networks.

International Routing Code A three-digit code within the North American Numbering Plan, beginning with 1, that classifies international calls as requiring either regular or special handling.

International Telecommunications Union (ITU) A European telecommunication standards body. Counterpart of ANSI.

Internet The major network running the Internet protocol across the United States and Canada. The Internet consists of more than 30,000 hosts and includes sites at universities, research laboratories, corporations, and nonprofit agencies.

Internet Service Provider (ISP) A vendor that provides access to the Internet and the World Wide Web.

Intranet A private network inside a company or organization that not only serves as the company information network but also provides access to the public Internet. The intranet appears like another server to the Internet.

Intersatellite Service Provides links between artificial earth satellites.

Inverse Multiplexing The process of splitting a high-speed channel into multiple signals and transmitting the multiple signals over multiple facilities operating at a lower rate than the original signal. It is followed by a process of recombining the separately transmitted portions into the original signal at the original rate of speed.

Local Access and Transport Area (LATA) As designated by the Modification of Final Judgment, an area in which a local exchange carrier is permitted to provide service. It contains one or more local exchange areas, usually areas with common social, economic, or other interests. (The

Modification of Final Judgment was the U.S. federal court decision to break up the AT&T Bell Telephone Company into seven Bell Operating Companies.)

Local Area Network (LAN) A private network offering high-speed digital communications channels for the connection of computers and related equipment in a limited geographic area. LANs use fiber optic, coaxial, or twisted pair cables or radio transceivers to transmit signals.

Local Exchange Carrier (LEC) A company that provides telecommunications service within a local access and transport area (LATA).

Local Loop A channel connecting customer equipment to the line-terminating equipment in a central office of the local exchange company. Typically, a loop is a two- or four-wire cable circuit between a vertical main distributing frame in a central office and the point of termination at a customer's premises. The cable from a frame to a cross-connect terminal is called a fraeler. The feeder can also be provided over digital loop or analog carrier systems. Digital systems can be served by either wire or fiber optics.

Message Waiting Notification (MWN) A service that informs enrolled subscribers when a voice message is available for retrieval. MWN may use a pip tone, an MS indication, or alert pip tone to inform a subscriber of an unretrieved voice message(s). MWN has no impact on a subscriber's ability to originate or receive calls.

Messaging Delivery Service (MDS) A service that permits pending voice messages to be attempted for delivery to a subscriber on a periodic basis until the subscriber acknowledges receipt of the messages.

Mobile Access Hunting (MAH) A call-processing service that causes a call to a pilot directory number to search a list of termination addresses sequentially for one that is idle and able to be alerted. If a particular termination address is busy, inactive, fails to respond to a paging request, or does not answer alerting before a time-out, then the next termination address in the list is tried. Only one termination address is alerted at a time. The mobile telephones in the group may be alerted using distinctive alerting.

Mobile Identification Number (MIN) The ten-digit number that represents a mobile telephone's (mobile station) identity.

Mobile Satellite Services (MSS) A communication service that provides communication between mobile earth stations and one or more space stations or between mobile earth stations via one or more space stations. Earth stations can be situated on ships, aircraft, and terrestrial vehicles. This service may also be used to detect and locate emergency signals from people in distress.

Mobile Switching Center (MSC) The central switching system used for cellular and PCS networks. The MSC was formerly called the mobile telephone switching office (MTSO).

Mobile Telephone Switching Office (MTSO) A cellular carrier switching system that includes switching equipment needed to interconnect mobile equipment with land telephone networks and associated data support equipment. *See also mobile switching center (MSC).*

Modem A contraction of modulator/demodulator. This is a device or circuit that converts digital signals to and from analog signals for transmission over conventional analog telephone lines. The term modem may also refer to a device or circuit that converts analog signals from one frequency band to another.

Multifrequency (MF) A type of in-band address signaling method in which decimal digits and auxiliary signals are represented by selecting a pair of frequencies from the following group: 700, 900, 1100, 1300, 1500, and 1700 Hz. These audio frequencies are used to indicate telephone address digits, precedence, control signals such as line-busy or trunk-busy signals, and other required signals.

Multiplexer Electronic equipment that allows two or more communications signals to pass over a single communications circuit.

Narrowband A communications channel of restricted bandwidth, often producing degradation of the transmitted signal.

National Number The telephone number identifying a calling subscriber station within an area designated by a country code.

Network (1. general) A series of points interconnected by communications channels, often on a switched basis. Networks are either common to all users or privately leased by a customer for some specific application. (2. antenna coupling) A network that employs a radio frequency circulator, enabling two separate radio transmitters to use the same antenna at the same time. (3. balanced) A network in which the series elements in both legs of the circuit are symmetrical with respect to ground. (4. bilateral) A network that employs signaling protocols enabling two or more carriers to communicate with one another as technical and business equals. SS7 is a signaling protocol that enables two or more carriers to maintain a bilateral (equal) business relationship.

Network Element A facility or the equipment used in the provision of a telecommunications service. The term includes subscriber numbers, databases, signaling systems, and information sufficient for billing and collection or used in the transmission, routing, or other provision of a telecommunications service.

Network Interconnection The interconnecting of two more networks.

Network Interface Card A circuit card or pack that connects a desktop computer to a larger network of computers. In telecom systems, network interface cards connect network devices to other network devices.

Network Operations Center (NOC) A center responsible for the surveillance and control of telecom traffic flow in a service area.

Noise (1. general) Any random disturbance or unwanted signal in a communication system that tends to obscure the clarity of a signal in relation to its intended use. (2. ambient) The acoustic noise that is part of the environment in which a system transducer is located. (3. atmospheric) A component of sky noise arising from natural phenomena within the atmosphere, such as a lightning discharge.

North American Numbering Plan (NANP) A telephone numbering system used in North America that uses ten-digit numbering. The number consists of a three-digit area code, a three-digit central office code, and a four-digit line number.

Number Portability The customer's current phone number can be transferred to other networks, carriers, or service providers.

Numbering Plan Area (NPA) A three-digit code that designates one of the numbering plan areas in the North American Numbering Plan for direct distance dialing. Originally, the format was NO/IX, where N is any digit 2 through 9 and X is any digit. From 1995 on, the acceptable format is NXX.

Off-Hook (1- telephony) A signal used on lines and trunks to indicate in-use or request-for-service states.

Offered Load The total traffic load submitted to a group of servers, including any load that results from retries.

On-Hook (1. telephony) A signal used to indicate that a line or trunk is not currently in use and is available for service. The signal is transmitted to a central office when the receiver is placed on its hook. (2. telephony) The state of being on hook.

Open Systems Interconnection (OSI) Model An internationally accepted framework of standards for telecommunications between different systems made by different vendors. The model organizes the telecom functions into seven different layers. It allows engineers to isolate and classify telecommunications into discrete functions or activities.

Operations Administration and Maintenance (OA&M) Managing a network or telecom system. Normally, one uses OA&M to refer to a series of functions that include: network diagnostics, network element management, alarm indications, and performance monitoring.

Operator Services Traditionally used by wireline telephone companies to refer to services that use a human or automated operator to render assistance to customers in the making of collect, third-party billed, credit card, calling card, and person-to-person calls.

Optical Fiber A threadlike filament of glass used to transmit digital voice, data, or video signals in the form of light pulses. Multimode step index fiber has relatively low capacity and is seldom used. Multimode graded index fiber is used for low-traffic routes. Singlemode step index fiber has high capacity and is the most commonly used type.

Out-of-Band Signaling A type of signaling in which the frequencies or timeslots used to carry the signals are outside the bandwidth of the information channel.

Outside Plant The part of a telephone system that is outside of local exchange company buildings. This includes cables and supporting structures. Microwave towers, antennas, and cable system repeaters are not considered to be outside plant equipment.

Packet Data Data transmission that breaks up or sends small packets of data by including the address and sequence number with each packet sent. Packets find their way through a packet switching network that eventually routes them to their destination, where they are placed back in sequence by a packet assembler/disassembler (PAD).

Packet Switching A mode of data transmission in which messages are broken into increments, or packets, each of which can be routed separately from a source; the packets can then be reassembled in the proper order at the destination.

Paging A method of delivering a message, via a public communications system or radio signal, to a person whose exact whereabouts are unknown by the sender of the message. Users typically carry a small paging receiver that displays a numeric or alphanumeric message on an electronic readout; the message could be sent and received as a voice message or other data.

Paging Audio Tone Signal Tone only, no alphanumeric characters at all. The tone means the individual should either call an operator or call a predetermined point. There are three types of tone-only protocols, of two, three, and five tones.

Paging Message Service (PMS) A service that permits attempts to deliver paging messages to a subscriber (via the SMS) on a periodic basis until the subscriber acknowledges receipt of the message. Messages may be up to 256 characters.

Password Call Acceptance (PCA) A call-screening feature that allows a subscriber to limit incoming calls to calling parties able to provide a valid PCA password (i.e., a series of digits).

Path Minimization The process of efficient fixed-network (the non-wireless portion of the call) routing of wireless call tables in the wireless carrier switches.

Permanent Virtual Circuit (PVC) A virtual circuit that provides the equivalent of a dedicated private-line service over a packet-switching network between two data terminal devices. The path between the users is fixed. However, a PVC uses a fixed logical channel to maintain a permanent association between the data terminal device stations.

Photonic Switches Switches that use light to carry telecommunications signals within the switch.

Point of Interface (POI) The physical location marking the point at which the local exchange carrier's service ends.

Point of Presence A physical location established by an interexchange carrier within a LATA for the purpose of gaining LATA access. The point of presence is usually a building that houses switching and/or transmission equipment, as well as the point of termination.

Preferred Language (PL) Allows a telephone service subscriber to choose service in English, Spanish, French or Portuguese as the language for network services. Future provision will be made for additional languages.

Presubscription A term created in 1984 to describe subscription to a predetermined Interexchange Carrier. The term is a legacy word from the days immediately after AT&T divestiture.

Primary Rate Interface In the Integrated Services Digital Network (ISDN), a channel that provides digital transmission capacity of up to 1.536 Mbit/s (1.984 Mbit/s in Europe) in each direction. The interface supports combinations of one 64-kbit/s D channel and several ~kbit/s B channels, or H-channel combinations.

Private Branch Exchange (PBX) A private switching system serving an organization, business, company, or agency, and usually located on a customer's premises.

Private Line Often used to connect the wireless carrier switch to a cell site or to connect two cell sites together. This connection may be two-wire or four-wire analog, DS-1, or DS-3 circuits.

Protocol (1. rules) A precise set of rules and a syntax that govern the accurate transfer of information. (2. connection) A procedure for connecting to a communications system to establish, carry out, and terminate communications.

Protocol Conversion The translation of the protocols of one system to those of another to enable communication by different types of equipment, such as data terminals and computers.

Provisioning The operations necessary to respond to service, trunk, and special-service circuit orders and to provide the logical and physical resources necessary to fill those orders.

Public Switched Telephone Network (PSTN) An unrestricted dialing telephone network available for public use. The network is an integrated system of transmission and switching facilities, signaling processors, and associated operations support systems that is shared by customers. PSTN is also called a public network, public-switched network, or public telephone network.

Radio Common Carrier A common carrier licensed by the Federal Communications Commission (FCC) to provide mobile telephone services using radio systems.

Radio Determination Satellite Service A service that uses radio signals from satellites to determine the position and velocity of an object.

Reconciliation Clearinghouse function for multiple carriers. Interconnected carriers terminate calls in each other's networks. Under such circumstances, the carriers are entitled of some form of compensation.

Regional Bell Operating Company (RBOC) One of the seven U.S. telephone companies, also known as Baby Bells, that resulted from the break up of AT&T.

Ring Network A data network of circular topology in which each node is connected to its neighbor to form an unbroken ring. A ring network in which one of the nodes exercises central control often is called a loop.

Router *See routing switcher.*

Routing Switcher (1. general) An electronic device that connects a user-supplied signal (audio, video, and/or data) from any input to any user-selected output. Inputs are called sources. Outputs are called destinations. (2. network) A device that forwards packets of a specific protocol type (such as P) from one logical network to another. These logical networks can be the same or different types.

Satellite (1. general) A body that revolves around another body of greater mass and has a motion primarily determined by the force of attraction of the more massive body. (2. communications) An orbiting space vehicle containing a set of transponders that receive signals from the ground and retransmit them to other ground-based receivers.

Satellite System A network of satellites and associated ground stations that transmit telephone, audio, video, and data signals between terrestrial points.

Selective Call Acceptance (SCA) (1. Belcore) A CLASS feature that enables a customer to receive calls selectively from previously specified telephone numbers. All other calls are intercepted by a recorded denial announcement or routed to an alternate directory number, depending on the subscriber's selection when the service is activated. CLASS is a service mark of Bellcore. (2. telephony) A call-screening service that allows a subscriber to receive incoming calls only from parties whose Calling Party Numbers (CPNs) are in an SCA screening list of specified CPNs.

Server (1. telecommunications network) The equipment or a call-carrying path that responds to a customer's attempt to use a network. (2. LAN) A processor that serves users on a local area network, for example, by storing and managing data files or by connecting users to an external network.

Service Access Code (SAC) The three-digit codes in the NPA (N 0/1 X) format that are used as the first three digits of a ten-digit address in a North American Numbering Plan dialing sequence. Although NPA codes are normally used for the purpose of identifying specific geographical areas, certain NPA codes have been allocated to identifying generic services or providing access capability; these are known as SACs. The common trait, in contrast to an NPA code, is that SACs are nongeographic.

Service Objective A statement of the quality of service to be provided to a customer.

Service Provider An generic name given to a company or organization that provides telecom service to customers (subscribers).

Short Message Service Point-to-Point (SMS-PP) Provides bearer service mechanisms for delivering a short message as a packet of data between two service users, known as Short Message Entities (SMEs). SMEs are SMS endpoints capable of composing or disposing of a short message. One or both of the service users may be a mobile station. The data packets are transferred transparently between two service users. The network or destination application generates negative acknowledgments when it is unable to deliver the message as desired.

Signaling Link A communication path that carries common channel signaling messages between two adjacent signaling nodes.

Signaling Point (SP) In the SS7 protocol, a node in a signaling network that originates and receives signaling messages, transfers signaling messages from one signaling link to another, or both.

Signaling System Number 7 (SS7) An out-of-band common-channel signaling protocol standard designed to be used over a variety of digital telecommunication switching networks. It is optimized to provide a reliable means for information transfer for call control, remote network management, and maintenance.

Signaling Transfer Point (STP) In a common-channel signaling network, a packet switch that uses the SS7 protocol to connect signaling links to network switching systems and other signaling transfer points.

SCP (Service Control Point) The database that contains subscriber profiles. SCPs can also provide assistance in the routing of a call.

SSP (Service Switching Point) The switch (or routing element) in the SS7 network. The switch would be a Class 5 central office in the wireline network (LEC). In the cellular or PCS world the SSP would be the wireless switch/mobile switching center.

Soft Handoff The reverse of the hard handoff scenario. Soft handoff is a "make-before-break" form of call handoff between radio channels, whereby the mobile handset temporarily communicates with both the serving cell site and the targeted cell site (one or more cell sites can be targeted) before being directed to release all but the final target cell site radio channel. Currently only CDMA supports soft handoff.

Space Operation Service Concerned exclusively with the operation of spacecraft.

Space Research Services A research service where spacecraft are used for scientific or technical research.

Star Network A data network with a radial topology in which a central control node is the point to which all other nodes join.

Supervisory Signals Signals used to indicate or control the status or operating states of circuits that establish a connection. A supervisory signal indicates that a particular state in a call has been reached and can signify the need for additional action.

Switched Circuits Circuits used in the support of wireless 911.

Switched Multimegabit Data Service (SMDS) A 1.544-Mbps public data service.

Switching (1. general) The process of making and breaking (connecting and disconnecting) two or more electric circuits. (2. telecommunications) The process of connecting appropriate lines and trunks to form a communications path between two or more stations. Functions include transmission, reception, monitoring, routing, and testing.

Synchronous (1. general) In step or in phase, as applied to two or more devices; a system in which all events occur in a predetermined timed sequence. (2. data communications) An operation that occurs at intervals directly related to a clock period. The bus protocol in such data transactions is controlled by a master clock and is completed within a fixed clock period.

Synchronous Optical Network (SONET) A standard format for transporting a wide range of digital telecom services over optical fiber. SONET is characterized by standard line rates, optical interfaces, and signal formats.

T Carrier Generic name given to describe any of the several analog transmission facilities used in the United States. T carriers were first used during the early 1960s and started off as a single twisted pairs of wires.

Tandem Switch A switch that supports a network topology in which connectivity between locations is attained by linking several locations together through a single point. The tandem switch is very similar to the "traffic cop," directing traffic from several streets into other groups of streets.

Tariffs Documents filed by a wireline telephone company with a state public utility commission or the FCC. A tariff describes in detail the company's services and pricing, essentially explaining and justifying the pricing of the service.

Telecommunications The transmission, between or among points specified by the user, of information of the user's choosing (including voice, data, image, graphics, and video), without change in the form or content of the information.

Telecommunications Carrier Any provider of telecom services. A telecom carrier is treated as a common carrier under the 1996 Act only to the extent that it is engaged in providing telecom services.

Telecommunications Service The offering of telecommunications for a fee directly to the public, or to such classes of users as to be effectively available directly to the public, regardless of the facilities used.

Telegraphy A form of telecommunications concerned with the process of providing transmission and reproduction at a distance of text material or fixed images. The transmission of such information may be by physical transmission facilities or over the air using some form of signaling protocol.

Television The transmission and reception of visual images via electromagnetic waves.

Terminal Equipment The computers, telephones, and other data or voice devices at the end of a telephone line.

Terminals Devices that typically provide the interface between the telecom system and the user. Terminals may be fixed (stationary) or mobile (portable).

Three-Way Calling (3WC) Provides the subscriber the capability of adding a third party to an established two-party call, so that all three parties may communicate in a three-way call.

Time Division Multiplexing (TDM) A method for sending two or more signals over a common transmission path by assigning the path sequentially to each signal, each assignment being for a discrete time interval. All channels of a time-division multiplex system use the same portion of the transmission links' frequency spectrum—but not at the same time. Each channel is sampled in a regular sequence by a multiplexer.

Toll Any message telecom charge for service provided beyond a local calling area.

Toll Center A pre-AT&T divestiture term, used to describe a Class 4 central office. This is also known as the tandem switching center.

Toll Free A term that refers to free calling numbers. A company would buy an "800" number to enable users to call it without any calling charges. Today this also includes any number users can call without incurring charges.

Traffic Engineering Planning activity that determines the number and type of communications paths required between switching points and the call-processing capacity of the switching equipment.

Translation (1. general) The conversion of information from one form to another. (2. switching system) The conversion of all or part of a telephone address destination code to routing instructions or routing digits.

Transmission Control Protocol/Internet Protocol (TCP/IP) The protocol that manages the bundling of outgoing data into packets, manages the transmission of packets on a network, and checks the bundles for errors. This networking protocol supports communication across interconnected information networks, i.e., the Internet.

Trunk A single transmission path connecting two switching systems. Trunks can be shared by many users, but serve only one call at a time.

Trunk Group A number of trunks that can be used interchangeably between two switching systems.

Trunk Side Connection Transmission facilities that connect switches to each other.

Twisted Pair A pair of insulated copper wires used in transmission circuits to provide bidirectional communications. The wires are twisted about one another to minimize electrical coupling with other circuits. Paired cable is made up of anywhere from a few to several thousand, twisted pairs.

Two-Wire Circuits A transmission circuit comprised of two wires that support the sending and reception of information.

Type 1 A trunk-side connection to an end office. The end office uses a trunk-side signaling protocol in conjunction with a feature known as Trunk with Line Treatment (TWLT).

Type 1 with ISDN An ANSI-standard ISDN line between the LEC end office and the MSC. The ISDN connection should be capable of providing connectivity to the PSTN that will support ISDN PRI (Primary Rate Interface) and ISDN BRI (Basic Rate Interface). The ISDN connection is a variation of the Type 1 connection.

Type 2A A trunk-side connection to the LEC's access tandem. It allows the MSC to interface with the access tandem as if it were an LEC end office. This connection enables the wireless carrier's subscribers to obtain presubscription. The service provider has access to any set of numbers within the LEC network.

Type 2B An interconnection type similar to the high-usage trunk groups established by the LEC for its own internal routing purposes. In the case of wireless carrier interconnection, Type 2B should be used in conjunction with Type 2A. When a Type 2B is used, the first choice of routing is through a Type 2B with overflow through the Type 2A.

Type 2C A telephone system interconnection intended to support interconnection to a public safety agency via an LEC E911 tandem or local tandem. It enables a wireless carrier to route calls through the PSTN to the Public Safety Answering Point (PSAP). This connection supports a limited capability to transport location coordinate information to the PSAP via the LEC.

Type 2D A telephone system interconnection intended to support interconnection to the LEC's operator service position. This connection enables the LEC to obtain ANI (Automatic Number Identification) information about the calling wireless subscriber in order to create a billing record.

Type S A telephone system interconnection that carries only control messages. The Type S is an SS7 signaling link from the wireless carrier to the LEC. The Type S supports call setup via the ISUP (ISDN User Part) portion of the SS7 signaling protocol and TCAP querying. Type S is used in conjunction with Types 2A, 2B, and 2D.

Universal Resource Locator (URL) Internet address used on the World Wide Web. A URL is a string expression that can represent any resource on the Internet.

Uplink The earth-to-satellite microwave link and related components such as earth-station transmitting equipment. The satellite contains an uplink receiver. Various uplink components in the earth station are involved with the processing and transmission of the signal to the satellite.

Validation Often confused with authentication. Where authentication essentially certifies the user as either "real" or "fake" and also as either "good" or "bad," validation certifies the "permission" to complete the call. Here is an example: A user calls another party using a mobile handset. The carrier certifies that the handset is the real one and is not a clone being used by a criminal. Once the handset has been given the "thumbs up," the carrier checks to see if the call is even allowed under the user's billing plan. This step is validation, being given the final "green light" to complete the call.

Very High Rate Digital Subscriber Line (VDSL) A high-speed transmission technology intended to be used in the local loop. VDSL technology supports fiber optics to a neighborhood while the last "mile" to the specific home still uses unshielded twisted pairs of wires.

Video An electrical signal that carries TV picture information.

Video On Demand (VOD) A service that telephone companies have been seeking to bring to their subscribers. VOD enables a subscriber to request any video he wishes to see.

Virtual Circuits A telecommunications link that appears to the user to be a dedicated point-to-point circuit.

Visitor Location Register (VLR) The part of a wireless network (typically cellular or PCS) that holds the subscription and other information about visiting subscribers authorized to use that network.

Voice Mail (VM) A computer-based system for the delivery, storage, forwarding, and retrieval of voice messages, especially those carried by telephone networks. Provides the subscriber with services, which include not only the basic voice recording functions but also time-of-day recording, and announcements, menu-driven voice recording functions, and time-of-day routing.

Wideband The passing or processing of a wide range of frequencies. Its meaning varies with context. In an audio system, wideband may mean a band up to 20 kHz wide; in a TV system, the term may refer to a band many megahertz wide.

Wink A telephone line signal that is a single supervisory pulse usually transmitted as an off-hook signal followed by an on-hook signal, where the off-hook signal is of a very short specified duration compared to the on-hook signal.

Wireless Local Loop (WLL) The providing of local telephone service via radio transmission.

xDSL A set of large-scale high-bandwidth data technologies that can use standard twisted pair copper wire to deliver high-speed digital services (up to 52 Mbps).

X.25 A data-link layer protocol; X.25 uses the LAPB portion of this layer.

APPENDIX B

BIBLIOGRAPHY

Chandler, Jr., Alfred D., *Strategy and Structure*, Anchor Books, Doubleday, New York, 1966.

Cummings, L.L., "Towards Organizational Behavior," Academy of Management Review, January 1978.

Dale, Ernest, "Planning and Developing the Company Organization Structure," Research Report 20, American Management Association, New York, 1952.

Duncan, Robert, "What is the Right Organization Structure," Organizational Dynamics, 1979.

Drucker, Peter, *Concept of the Organization*. John Day, New York 1946.

Engineering and Operations in the Bell System, AT&T, 1982.

Harte, Lawrence, *Dual Mode Cellular*, Steiner Publishing, 1992.

Louis, P.J., *Telecommunications Internetworking*, McGraw-Hill, New York, 2000.

Louis, P.J. *M-Commerce Crash Course*, McGraw-Hill; New York, 2001.

Notes on the Network 1980, AT&T, 1980.

Perrow, Charles, *Organizational Analysis: A Sociological View*, Wadsworth, California, 1970.

Telecommunication Transmission Engineering, AT&T, 1977.

Young, Harry (1992-1999), consolidated work of Harry Young, network interconnection consultant pioneer.

INDEX

Note: Boldface numbers indicate illustrations.

ABOUT THE AUTHOR

P. J. Louis has nearly a quarter of a century's worth of experience in the telecom business. He is currently a director at PricewaterhouseCoopers, LLP, in the Financial Advisory Services area. Prior positions include Vice President of Carrier Marketing & Product Management at TruePosition, Inc., a leading provider of wireless location services. Mr. Louis has also served as chief of staff for engineering at NYNEX (today known as Verizon). He has held a number of leadership positions within Bell Communications Research and NextWave Wireless. Mr. Louis is a former officer of the Institute of Electrical and Electronics Engineers (IEEE) Communications Society—New York Section. He has been a senior member of IEEE for 25 years. Mr. Louis' experience includes leading sales and marketing as well as technology organizations. He has been a featured speaker at telecommunications conferences and is the author of three other McGraw-Hill books: *M-Commerce Crash Course*, *Broadband Crash Course*, and *Telecommunications Internetworking*.